The Jahn-Teller Effect in C_{60} and
Other Icosahedral Complexes

The Jahn-Teller Effect in C_{60} and Other Icosahedral Complexes

C. C. CHANCEY and M. C. M. O'BRIEN

PRINCETON UNIVERSITY PRESS

PRINCETON, NEW JERSEY

Library of Congress Cataloging-in-Publication Data
Chancey, C. C.
The Jahn-Teller effect in C_{60} and other icosahedral complexes /
C. C. Chancey and M. C. M. O'Brien.
p.   cm.
Includes bibliographical references and index.
ISBN 0-691-04445-7 (cloth : alk. paper)
1. Electronic structure. 2. Buckminsterfullerene—Spectra.
3. Fullerenes—Spectra. 4. Jahn-Teller effect. 5. Icosahedra.
I. O'Brien, M. C. M. II. Title.
QC176.8.E4C52 1997   97-9078
546'.681—dc21   CIP

Contents

Figures

Tables

Glossary of Symbols

$\oplus$	Direct sum, as of electronic spaces or vibrational spaces.
$\otimes$	Direct product, as between electronic and vibrational spaces.
$\lvert\Gamma\rvert$	The dimensionality of irrep Γ.
$[\Gamma \oplus ..]_S$	Symmetric part of the Kronecker square of irreps.
$\{\Gamma \oplus ..\}_A$	Antisymmetric part of the Kronecker square of irreps.
a	Phonon annihilation operator.
$a^\dagger$	Phonon creation operator.
$\mathbf{a}_\gamma^\Gamma$	Matrix of electronic bases at minima of $\Gamma \otimes \gamma$ JT system.
A_g, A_u	Symmetric (g) or antisymmetric (u) one-dimensional irrep.
$A_D(\alpha)$	α-dependent rotation matrix in the h vibrational coordinate space.
APES	Adiabatic Potential Energy Surface.
$B(\gamma)$	γ-dependent rotation matrix in the h vibrational coordinate space.
β	Vibrational singlet irrep.
$C(\theta)$	θ-dependent rotation matrix in the h vibrational coordinate space.
C_{60}^*	Neutral C_{60} in an excited electronic state.
$^{(3,1)}C_{60}$	Neutral C_{60} in a (triplet, singlet) spin state.
$C_{60}^{n\pm}$	C_{60} in a charged state.
$D(\phi)$	ϕ-dependent rotation matrix in the h vibrational coordinate space.
D_{5d}	One of the Dihedral groups; represents a distortion symmetry of C_{60}.
D_{3d}	One of the Dihedral groups; represents a distortion symmetry of C_{60}.
D_{2h}	One of the Dihedral groups; represents a distortion symmetry of C_{60}.
Δ	Jahn-Teller induced or other energy splitting between states.
E	Electronic doublet irrep label.
E_{JT}	Jahn-Teller stabilization energy; reduction in system energy due to linear Jahn-Teller coupling.
ϵ or e	Vibrational doublet irrep in cubic symmetry.
G	Electronic quartet irrep label.
g	Vibrational quartet irrep label.
g_i	Basis function for the G space or a g-type phonon coordinate, with $i = 1, \ldots, 4$.
g	g-factor for a spin Hamiltonian.
$\overline{\mathbf{g}}$	g-tensor for a spin Hamiltonian.
γ, θ, ϕ	Euler angle for a rotation.
Γ	Electronic irrep label, with $\Gamma = T, G, H$.

Γ_i	For $i = 6, \ldots, 9$, irreps of the double icosahedral group.
γ	Vibrational irrep label, with $\gamma = g, h$.
$\Gamma \otimes \gamma$	Jahn-Teller coupling between a Γ electronic state and a set of γ-type vibrations.
$\mathcal{H}$	Hamiltonian operator.
H	Electronic quintet irrep label.
h_α	Vibrational quintet irrep label, with $\alpha \, (= 1, 2, a, b)$ a matrix representation label; $\alpha = 2$ for the $L = 2$ matrix, and $\alpha = 4$ for the $L = 4$ matrix.
h_i	Basis function for the H space or an h-type phonon coordinate, with $i = 1, \ldots, 5$.
HOMO	Highest Occupied Molecular Orbit.
I	Icosahedral group.
I_h	Full icosahedral group: I plus the parity operation, P.
I_2, I_3, I_3'	Icosahedral invariants.
J	Angular momentum operator.
$K(\Gamma)$	Ham reduction factor for an operator labelled by irrep Γ.
k_{eff}	Coupling constant in the cluster approximation.
k_γ^Γ	Coupling constant for a linear interaction between an electronic Γ state and a vibrational γ state.
L	Orbital angular momentum operator.
L	Orbital angular momentum quantum number.
$L_{ij}^{(2)}$	Second-order tensor operator $L_i L_j + L_j L_i$.
LUMO	Lowest Unoccupied Molecular Orbit.
$\overline{\Lambda}$	Tensor of expectation values of $L_{ij}^{(2)}$ in a state.
λ	Spin-orbit interaction coupling strength parameter.
$\lambda_x, \lambda_y, \lambda_z$	Cartesian components of a vibrational space angular momentum operator.
MCD	Magnetic circular dichroism (in absorption spectra).
$M^\Gamma(\gamma)$	Linear Jahn-Teller interaction matrix between the electronic states labelled by Γ and the vibrational modes labelled by γ.
ω_{eff}	Effective frequency in the cluster approximation.
p_i	Basis function for the T_{1u} ($L = 1$) space, with $i = x, y, z$.
p^n	Multiply-occupied T_{1u} electronic state with n electrons.
P	The parity operation, one of the elements in the full icosahedral group I_h.
Φ	Angular part of the wave function.
$\psi(Q)$	Vibrational wave function.
$\Psi(\mathbf{r}, Q)$	Born-Oppenheimer product of vibrational and electronic wave functions.
Q_γ^Γ	Matrix of vibrational coordinates at minima of $\Gamma \otimes \gamma$ JT system.
Q_i	Vibrational coordinates in the $T_1 \otimes h$ interaction.

$\mathbf{S}$	Spin angular momentum operator.
$S_{ij}^{(2)}$	Second-order tensor operator $S_i S_j + S_j S_i$.
S	Overlap of two localized wave functions.
SO(n)	Group of rotations in n-dimensional space.
T	Electronic triplet irrep in cubic symmetry.
T_h	Tetrahedral group.
T_i	Electronic triplet irrep labels for $i = 1, 2$ (labeling distinct matrix representations).
t_i	Basis function for the T_1 or T_2 spaces, with $i = 1, 2, 3$.
$\mathcal{T}$	General rotation in the h vibrational coordinate space.
τ_g or t_2	Vibrational triplet irrep in cubic symmetry.
$\mathbf{U}^\Gamma(\Lambda\,\lambda)$	Symmetric coupling matrix for an interaction between an electronic Γ state and a vibrational Λ state.
$u_n(\mathbf{r}, Q)$	Electronic wave function in a Born-Oppenheimer product.
V_{icos}	Icosahedral invariant in a spherical subspace.

The Jahn-Teller Effect in C_{60} and Other Icosahedral Complexes

1

Introduction

1.1 FROM PAST TO PRESENT

New interest in vibronic coupling and the Jahn-Teller effect has followed on several recent discoveries in physics and chemistry: high-temperature superconductivity, buckminsterfullerene (C_{60}), and the Berry phase all provide examples of vibronic, Jahn-Teller-active systems. These systems add to the list that has been growing since Bleaney and Bowers (1952) provided the first clear experimental evidence of a Jahn-Teller system. Early work, both experimental and theoretical, concentrated on systems possessing symmetries that were octahedral or lower. The stimulus for this research was the wide interest in point defects, particularly transition metal ion impurities, during the 1960s and 70s. Theoretical work on icosahedral Jahn-Teller systems was sparse, principally due to the rarity of physical examples. The 1985 discovery of the C_{60} molecule, which is Jahn-Teller active when ionized, galvanized interest in such systems (Kroto et al., 1985), and other Jahn-Teller icosahedral systems have followed, such as Si_{13}^{+} and other atomic clusters (Gu, Li, and Zhu, 1993; Röthlisberger et al., 1992). Almost all of the electronic and vibrational states of C_{60} are highly degenerate due to the high symmetry of the icosahedral group (I_h). The Jahn-Teller interaction, because it involves cases of electronic degeneracy, is thus a fundamental consideration in unraveling the structural and electronic properties of this and other icosahedral complexes. The range of Jahn-Teller interactions within the various I_h irreducible representations adds a new and rich domain to the older ongoing areas of Jahn-Teller research.

In introducing current aspects of the Jahn-Teller effect, we first turn to a brief historical outline of its antecedents, followed by an illustrative example and a short discussion of the Berry phase. Chapter 2 discusses physical systems that possess icosahedral symmetry and provides a summary of the aspects of group theory that we will make use of, including Ham reduction factors. The $T \otimes h$ system, often used to model the superconducting fullerides, is studied in Chapter 3. Chapter 4 deals with those complexes involving an electronic quartet: the so-called $G \otimes g$, $G \otimes h$, and $G \otimes (g \oplus h)$ Jahn-Teller systems. Those systems involving electronic quintets ($H \otimes g$, $H \otimes h_2$, and $H \otimes (g \oplus h)$) are studied in Chapter 5. Finally, the problems of electron spin resonance, multiple phonon modes, and molecular spectra are discussed in Chapter 6.

1.2 DEFINITION AND HISTORY

The Jahn-Teller effect is concerned with one aspect of the fundamental problem of how to describe quantum systems in which electronic and nuclear degrees of freedom are coupled. It is an example of electron-phonon coupling that is very simple to understand in its primitive form and yet produces a rich variety of phenomena to study. The use of the term electron-phonon implies that there are both heavy and light particles to be considered, with the motion of the heavy particles (ions) being discussed in terms of normal coordinates or phonons. The peculiarity of the Jahn-Teller interaction is that there must be a multiplicity of electronic states interacting with one or more normal modes of vibration. The Jahn-Teller theorem says that for almost any set of degenerate electronic states associated with a molecular configuration, there will exist some symmetry-breaking interaction in which molecular distortion is associated with the removal of the electronic degeneracy. The first full explanation of the effect was given by Hermann Jahn and Edward Teller in the spring of 1936, at the Washington meeting of the American Physical Society (Jahn and Teller, 1936). However, the roots of a theoretical understanding of the effect can be found two years earlier in Copenhagen, during overlapping visits by Teller and Lev Landau to the institute of Niels Bohr (Teller, 1972, 1982). At that time Teller and Landau had several discussions concerning degenerate electronic states in linear molecules such as CO_2. Landau's intuition was that a molecule in an orbitally degenerate electronic state would be inherently unstable with respect to symmetry-lowering distortions of its nuclear configuration. Teller managed to convince Landau that linear molecules were an exception to this general supposition. In doing this Teller was able to rely on the work of Rudolph Renner, a recent Ph.D. student of his at Göttingen, whose thesis had dealt with linear triatomic molecules (Renner, 1934). The following year while in London, Teller returned to the topic and addressed the question of whether there existed any other exceptions to Landau's hypothesis. In this he was joined by Jahn, and together they demonstrated that linear molecules were the sole exception in cases of orbital degeneracy. The other exception, Kramers degeneracy, cannot be lifted through any nuclear displacement, since it is due to time reversal invariance of the Hamiltonian (Jahn, 1938).

The simplest proof of the Jahn-Teller theorem rests on group theory: one lists all the possible point symmetry groups under which a system such as a molecule may be invariant, and one also lists all the normal modes of each system, classified by their symmetry. Since all electronic states in such a system can also be classified by symmetry, and each labeled by an irreducible representation (irrep) of the point group for the high-symmetry ionic configuration, one can show that a linear electron-phonon interaction that lowers ionic symmetry is *permitted* by considerations of symmetry in almost every case. The exceptions are linear molecules and molecules in Kramers-degenerate electronic states.

This is as far as the proof goes—the interaction is allowed—but we normally make the usual assumption that anything allowed will be disposed to occur. The ionic distortions capable of lifting the degeneracy can also be classified by symmetry, using irreducible representations, and group theory tells us which these are. For spin-independent electron systems, the irreducible representation symmetry label for the distortion must occur in the symmetric square formed using the electronic state's irreducible representation. The ionic distortions, or phonon modes, defined in this way are called the *Jahn-Teller-active* modes. (Details and further discussions of the proof are available in the general reviews by Englman [1972] and by Bersuker and Polinger [1989].)

1.3 THE $E \otimes \beta_1$ INTERACTION

The simplest example of the Jahn-Teller effect is that of a doubly degenerate electronic state interacting with a single mode of vibration. All such interactions can be reduced to the same form of Hamiltonian, and so this system occurs widely. For instance, most Jahn-Teller interactions responsible for structural phase changes are modelled in this way. The type of Jahn-Teller interaction depends on the symmetry of the complex in which it takes place, and this degree of simplicity will be expected if there is a relatively low symmetry to start with. Though this system will not appear in icosahedral complexes, it provides an excellent introduction to aspects of Jahn-Teller systems that will appear again in more complex form in later chapters, and we discuss it here for that reason. The Hamiltonian can be written

$$\mathcal{H} = -\frac{1}{2}\frac{\partial^2}{\partial Q^2} + \frac{1}{2}Q^2 + k\begin{bmatrix} -Q & 0 \\ 0 & Q \end{bmatrix}, \tag{1.1}$$

where Q is the normal mode coordinate and the matrix operates within the pair of electronic states as basis. The energy eigenvalues of this Hamiltonian will be in units of $\hbar\omega$, where ω is the normal mode frequency of oscillation. In coming chapters, we shall usually arrange to have energies measured in these terms, and exceptions will be explicitly noted. The assumption of a single effective mode, ω, is well justified for many physical systems. This Hamiltonian can serve to represent *all* the possible two-by-one Jahn-Teller interactions, since all can be transformed into this form in which all masses and vibrational frequencies have been transformed out. The only parameter left in, k, tells us how strong the Jahn-Teller coupling is in comparison with the separation of the vibrational energy levels, $\hbar\omega$. It is because so many systems can be reduced to simple Hamiltonians like this one, of which there are relatively few, that discussions of the Jahn-Teller effect usually concentrate on working out the implications of such Hamiltonians. This simple Hamiltonian has a simple set of solutions. It

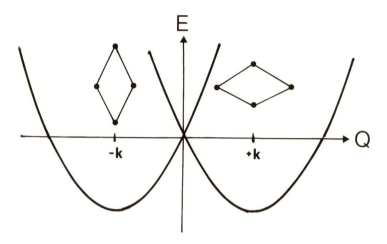

Figure 1.1. The adiabatic potential energy "surfaces" for the simple E ⊗ β_1 Jahn-Teller system. For a simple square planar molecule, each APES minimum corresponds to the molecular distortion shown.

can be rewritten as a pair of Hamiltonians:

$$\mathcal{H} = -\frac{1}{2}\frac{\partial^2}{\partial Q^2} + \frac{1}{2}Q^2 \pm kQ. \qquad (1.2)$$

These represent a pair of harmonic oscillators with origins displaced a distance $\pm k$ from $Q = 0$ and potential minima of $E = -k^2/2$. The potentials for such a pair of oscillators are shown in Figure 1.1. A simple square-planar molecule offers a physical setting for an E⊗β Jahn-Teller interaction. Figure 1.2 shows the physical distortion modes that such a system undergoes. Although it is quite simple, this picture can be used to illustrate some important ideas. First we see the effect of the size of parameter k. If k is large, by which we mean $k \gg 1$, then the lowest energy levels are localized in the bottom of the two wells with very little overlap. The ground state of the combined system is still doubly degenerate, but there is little overlap between its two components, and for all practical purposes there are two alternative, equally probable, ground states corresponding to the two distortions $Q \rightarrow Q \pm k$. These distorted states are of lower symmetry than the uncoupled states, and it is in this sense that the Jahn-Teller interaction is symmetry-breaking. It is sometimes said that this interaction also lifts the degeneracy, but it is clear that the two fold degeneracy of the ground state still exists even though the two components are disjoint. If, on the other hand, the coupling is weak, as when $k \leq 1$, then there is a substantial overlap between the two ground states and any attempt to measure a physical lowering of site symmetry will produce a result that averages out

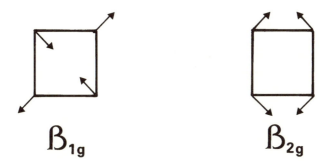

Figure 1.2. Distortion modes for a square-planar molecule under the influence of an $E \otimes \beta$ Jahn-Teller effect. The β_{1g} mode is the mode in (1.1), which is coupled to an electronic doublet. The other vibration represents the other β normal mode of a square-planar molecule: B_{2g}, a mode altering bond angles.

the effect of the Jahn-Teller distortion. Yet vibronic effects can be observed even in weak coupling: the coupling of vibrational states into electronic states reduces the strength of electronic interactions (such as spin-orbit coupling). This occurence, known as the Ham effect, signals the involvement of Jahn-Teller interactions and is prime evidence of their effects. We will return to a discussion of the Ham effect and Ham reduction factors in Chapter 2. The first of these cases ($k \gg 1$) is termed the *static* Jahn-Teller effect; the other ($k \leq 1$) represents what is called the *dynamic* Jahn-Teller effect. Static effects can be seen in crystal structure, in spin resonance, in nuclear magnetic resonance (NMR), or in the Mössbauer effect. Such distortions also may be stabilized by effects such as strains in a crystal. Whether a static effect is observed depends in part on the interrogation time of the experimental technique used. As Sturge (1967) has emphasized, the same system can, in principle, show a static effect in an experiment with a short characteristic time (such as spin resonance), and yet show no distortion because of the dynamic effect when studied in an experiment with a relatively long characteristic time (such as NMR).

1.4 THE BERRY PHASE

The term "Berry phase" will occur frequently throughout this book. It refers to a phase acquired by a quantum system moving adiabatically around a circuit in the parameter space of the system. It was first introduced by Berry in 1984 and is a concept that has thrown light on a wide variety of phenomena.[1] The starting point in understanding the concept begins with the observation that

---

[1] A useful general account of the Berry phase has been given by Aitchison (1988).

most physical systems can be divided into "fast" and "slow" components. A molecule is an example, since it is composed of fast (light) outer electrons and slow (heavy) atomic cores. The adiabatic approximation (Appendix A) uses the difference in the "fast" and "slow" energy scales to define an approximate molecular wave function, one that is a Born-Oppenheimer product of what are essentially electronic and atomic core wave functions. Berry's insight was to demonstrate that these two parts of the total wave function are coupled together in a simple geometrical way. The evidence of this coupling is the appearance of a phase factor in the molecular wave function (the Berry phase factor). The phase in question, denoted by $\gamma(c)$, is the integrated change in phase of the wave function of the fast component when the system is carried round a closed curve, c, in the coordinate space of the slow component. For an example, we look at a typical wave function in the adiabatic approximation, one term of (A.4),

$$\Psi_n(\mathbf{r}, Q) = \psi_n(Q) u_n(\mathbf{r}, Q), \tag{1.3}$$

where Q represents the slow coordinates of the atomic cores and $\mathbf{r}$ is the set of fast coordinates of the outer electrons. In this approximation the electronic wave function, $u_n(\mathbf{r}, Q)$, depends parametrically on the Q's. This dependence on the "slow" coordinates means that the "fast" component wave function, $u_n(\mathbf{r}, Q)$, can only combine with the slow component wave function, $\psi_n(Q)$, in certain ways. As the atomic cores oscillate in a linear combination of molecular normal modes, the molecule executes a closed path in Q space (one cycle per normal mode oscillation). The parametric coupling between Q and $u_n(\mathbf{r}, Q)$ means that there must be a coordinated change in the wave function phases of both ψ and u_n. Berry showed that a phase change of π could be expected to take place if the circuit enclosed a point where states were degenerate, while if no degeneracy or an even number of degeneracies was enclosed, the phase change would be zero. In quoting these results, it should be remembered that Berry's analysis was restricted to parameter spaces of two or three dimensions. Generalization to higher-dimension spaces will require careful consideration; this topic will be addressed further, as appropriate, in coming chapters.

The most direct way of calculating the Berry phase follows from the requirement that Ψ be single-valued in the space of the Q. In the case of a Jahn-Teller system, the electronic wave function, u_n, emerges as the eigenvector of a matrix that depends on Q at each point on the path. Because of this, u_n can be multiplied by any phase factor at each point, and this phase factor can then be chosen so that u_n itself satisfies the above single-valuedness criterion. Now the special feature of all Jahn-Teller matrices is that they are real, and thus the phase can always be chosen so that u_n is real. This means that in taking u_n continuously around a closed circuit, it can be assumed to be real, and there can only be a phase change of π or zero, corresponding to a change of sign or not. If the sign is unchanged u_n is single-valued. If this is not the case, then u_n must be made single-valued by multiplying it by an appropriate phase factor of the form $\exp(i\phi)$ such that ϕ changes by π round the circuit. We thus see that $\gamma(c)$ can be found equally well by tracking the sign of the real wave function u_n round the path c. In applying the concept of Berry's phase to Jahn-Teller and other adiabatic systems, the quantity that has real importance is the product $\Psi_n = u_n(\mathbf{r}, Q)\psi_n(Q)$. This must be single-valued in Q space to be well defined. With u_n real, the adiabatic approximation gives us a real $\psi_n(Q)$, which thus has to change sign round a closed circuit if u_n does so. As a result, its phase is the same as the phase of the real u_n. This is the aspect of the Berry phase that is useful in studying the Jahn-Teller states, and it is often of crucial importance in determining the nature of the eigenstates.

1.5 ICOSAHEDRAL COMPLEXES

The anions (and cations) of C_{60} and various icosahedral atomic clusters are Jahn-Teller active. The evidence for this is both theoretical and experimental: structural and molecular dynamics calculations and spectral analyses are in agreement that Jahn-Teller interactions play vital roles in determining ground state properties in these icosahedral complexes (Pickett, 1994; Röthlisberger et al., 1992; Gu, Li, and Zhu, 1993). The abstract Hamiltonians studied in Chapters 3–5 form the theoretical foundation for a better understanding of these real-world systems, and we briefly review these systems from the perspective of their Jahn-Teller activity before considering icosahedral symmetry in greater detail in Chapter 2. We will discuss experimental aspects of several of these physical systems in more detail in Chapter 6.

1.5.1 C_{60}–Based Materials

Hückel molecular orbital (HMO) theory provides a useful starting perspective for considering C_{60} and its anionic and cationic forms. Figure 1.3 shows the HMO energy levels for C_{60}. In its ground state, neutral C_{60} possesses a closed shell electronic structure and no Jahn-Teller interaction is expected. However, because the lowest unoccupied molecular orbital (LUMO) is threefold degenerate, a Jahn-Teller interaction is possible when C_{60} is excited and also for anions such as C_{60}^- and for C_{60}^{3-} (the charge state in alkali-doped solids such as Rb_3C_{60}) (de Coulon et al., 1992; Negri et al., 1988). Cations such as C_{60}^+ also will be Jahn-Teller active, since the highest occupied molecular orbital (HOMO) for C_{60} is fivefold degenerate. Given that the molecular orbitals will be delocalized over a relatively large surface area (C_{60} is roughly 8 Å in diameter), it is unlikely that any Jahn-Teller distortions will be large. However, C_{60} does distort along the h_g Jahn-Teller-active coordinates upon being excited into one of its low-lying triplet orbitals (either the LUMO or the next lowest, which is also threefold degenerate). Negri et al. (1988) have calculated the Jahn-Teller

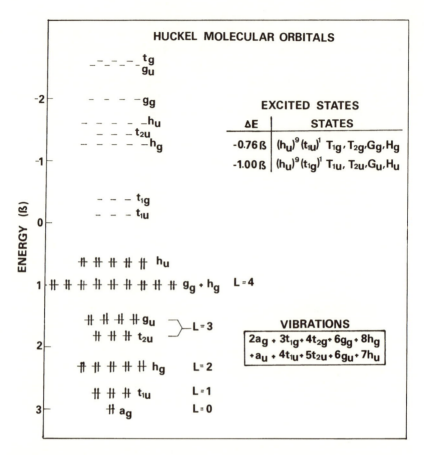

Figure 1.3. Hückel molecular orbital energy levels for C_{60}. The neutral molecule possesses a closed shell (ten electrons in an electronic quintet level labelled by h_u). The two lowest excited configurations are each fifteenfold degenerate, $(h_u)^9(t_{1u})^1$ and $(h_u)^9(t_{1g})^1$. These configurations yield eight excited states, each of which is at least threefold degenerate. *Source*: Haddon et al. (1986).

distortions for these cases and find their results in good agreement with spectroscopic data. Their quantum chemical calculations show that the h_g mode near 260 cm^{-1} is the mode most strongly involved of the Jahn-Teller modes available. This mode, an oblate spheroidal or "squash" mode, can be pictured by setting the C_{60} on a surface with a pentagonal face down, then applying pressure on top, on the pentagonal face opposite the first (Pickett, 1994). The Jahn-Teller activities of the negatively charged species of C_{60}, such as C_{60}^- and C_{60}^{3-}, are of interest due to the possibility of an electron-phonon source for the fullerides' superconductivity (Pickett, 1991). When electrons are added to C_{60}, they begin to fill the triply degenerate orbital, previously the LUMO for neutral C_{60}. (This picture of orbital filling is useful even for solid C_{60} compounds: solid C_{60} is weakly bound by van der Waals forces, and intermolecular interactions should be an order of magnitude smaller than the intramolecular energies, an assumption in keeping with the molecular dynamics studies of solid C_{60}, which show that individual molecules rotate freely at temperatures above 260 K [Tycko et al., 1991].) The Jahn-Teller interactions within the C_{60}^- and C_{60}^{3-} anions thus are of considerable interest, and the Jahn-Teller case to be considered is again that of an orbital triplet interacting with a fivefold h_g distortion mode. The interaction of a triplet with an h_g vibrational mode has been analyzed by Lannoo et al. (1991), who draw on the earlier work of Öpik and Pryce (1957) and O'Brien (1969). Lannoo et al. estimate that the intra-C_{60} electron-phonon coupling arising from Jahn-Teller interactions is sufficiently strong to account for the observed T_c values in the fullerides. In related work, de Coulon et al. (1992) have estimated a lower bound of 24 meV for the Jahn-Teller distortion energy of C_{60}^-. In calculating this, de Coulon et al. consider an h_g distortion that is symmetric with respect to one of the icosahedron's fivefold symmetry axes. Such a distortion breaks the icosahedral symmetry in a minimal way yet is capable of stabilizing the added electron in an equatorial orbital on the molecule (Pickett, 1991). This distortion symmetry is in keeping with the results of Negri et al. (1988), who also have calculated the Jahn-Teller distortions in C_{60}^-.

1.5.2 Other Icosahedral Clusters

As Fowler and Manolopoulos (1995) have recently outlined, C_{60} is merely the most prevalent of a larger set of fullerene clusters with icosahedral symmetry. In general, the Coxeter tesselation method produces an icosahedral fullerene structure C_n for any $n = 20(a^2 + ab + b^2)$, where $b \geq 0$ and $a \geq b$ with a and b both integers (Coxeter, 1971; Fowler, Cremona, and Steer, 1988). C_{20}, C_{80}, and C_{240} all possess, or are expected to possess, icosahedral symmetry (in the absence of any Jahn-Teller interactions). Table 1.1 shows the ground-state electronic structures of C_{20} and C_{80} (both observed experimentally) and the putative icosahedral fullerenes C_{180}, and C_{240} (Fowler and Manolopoulos, 1995). Of these, C_{20} is known to have a strong Jahn-Teller interaction, leading to static deformation and lowering of symmetry from I_h to C_2 (Zhang et al., 1992).

Other examples arise in boron-rich solids, semiconducting materials that have unique three-centered covalent bonds, and crystalline structures with icosahedral B_{12} or soccer-ball B_{84} clusters (Kimura 1993). Kimura et al. (1993) have performed molecular orbital calculations for a $B_{12}H_{12}$ icosahedral cluster that predict a Jahn-Teller instability of cubic or rhombohedral symmetry. The highest molecular orbital of neutral $B_{12}H_{12}$ has a fourfold degeneracy and only

TABLE 1.1

Other Icosahedral Fullerenes

The electronic configurations are determined by assuming that there is one valence π electron per carbon atom and then applying Hund's Rules. The S, L, and J values are those for the ground state, assuming Hund's Rules. For icosahedral symmetry, the ground state J decomposes into a direct sum of the I_h irreps given below.

Complex	Electronic configuration	S	L	J	Ground state irreps (I_h)
C_{20}	$s^2 p^6 d^{10} f^2$	1	5	4	G_g, H_g
C_{80}	$s^2 p^6 d^{10} \ldots h^{22} i^8$	4	20	16	$A_g, 2T_{1g}, T_{2g}, 2G_g, 3H_g$
C_{180}	$s^2 p^6 d^{10} \ldots l^{34} m^{18}$	9	9	0	A_g
C_{240}	$s^2 p^6 d^{10} \ldots m^{38} n^{40}$	1	19	20	$A_g, 2T_{1g}, 2T_{2g}, 2G_g, 4H_g$

Source: Dresselhaus, Dresselhaus, and Saito (1993).

six electrons, thus being a possible example of a quartet Jahn-Teller system ($G \otimes (h \oplus g)$). Icosahedral structures of 13, 19, 55, and 147 atoms have been experimentally observed by Echt et al. (1981) in xenon clusters, and icosahedral packing appears to dominate nickel clusters in the range of 49 to 105 atoms (Parks et al., 1991). The best-studied cases of icosahedral clusters (excluding C_{60}) are those composed of 13 atoms. These, the smallest icosahedral structures, provide a wealth of possible Jahn-Teller active systems: the HOMO's of Si_{13}, Na_{13}, Mg_{13}, and Al_{13} are all partially filled and degenerate for icosahedral symmetry, promoting Jahn-Teller instabilities in each case (Röthlisberger et al., 1992). The HOMO for the Si_{13}^+ cluster is a fourfold degenerate G_g orbital with one-electron occupation, and thus should also be Jahn-Teller active (Gu, Li, and Zhu, 1993). Al_{13} has a HOMO that is threefold degenerate with 5/6 occupation, and it offers a physical check for abstract $T \otimes h$ Jahn-Teller models. Other cluster systems, such as Si_{12}, may provide physical representations for Jahn-Teller models of electronic quintets in icosahedral surroundings (Gu, Li, and Zhu, 1993).

2

Icosahedral Symmetry and Its Effects

2.1 ICOSAHEDRAL SYMMETRY

Icosahedral symmetry is rare in nature, though examples such as C_{60} and other fullerene molecules have become increasingly evident in recent years. The relevant point group I_h, with 120 distinct symmetry elements, is the largest of the 32 point groups. This high degree of symmetry means that icosahedral complexes will be highly degenerate in their vibrational levels and electronic states, producing a rich field for vibronic interactions. That icosahedral symmetry can be central to a physical description is seen in C_{60}: the Raman spectra for solid C_{60} (tetrahedral symmetry) and for alkali-metal-doped C_{60} compounds (cubic site symmetry) are surprisingly similar to the Raman spectra for the isolated C_{60} molecule, confirming the importance of the icosahedral structure as the natural starting point for perturbative methods. Our discussion of icosahedral symmetry will be outlined in terms of C_{60}, the best known icosahedral complex. However, the discussion and results—especially in later chapters—apply equally well to other icosahedral molecules, whether other fullerenes (such as C_{20} or C_{80}) or atomic clusters (such as Si_{13}^{+}).

2.2 GEOMETRY OF C_{60} AND THE GROUP I_h

Buckminsterfullerene, C_{60}, possesses the geometry of a truncated regular icosahedron (Fig. 2.1). A picture of this geometry can be built up by beginning with a regular icosahedron: a closed surface, constructed of twenty equilateral triangles, possessing twelve vertices (each a meeting point for five triangles). If lines are drawn from this figure's center through each of the twelve vertices, truncation occurs along planes perpendicular to these lines. Each plane will intersect the regular icosahedron to form a pentagonal face, producing a figure with twelve pentagonal faces. The truncation process also produces twenty hexagonal faces—the remains of the twenty triangular faces of the regular icosahedron. The symmetry of C_{60} is realized when each truncation creates twenty regular hexagonal faces, as shown in Figure 2.1. In the C_{60} molecule, a carbon atom occupies each vertex of the truncated regular icosahedron (Figure 2.2). Simple chemical analysis predicts that the two types of edges in the truncated

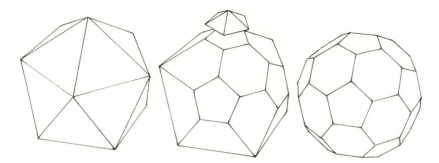

Figure 2.1. Producing a regular truncated icosahedron: (left) the regular icosahedron; (center) creating a pentagonal face by truncating a vertex; (right) the geometry of C_{60}, a regular truncated icosahedron. *Source*: Chung and Sternberg (1995).

icosahedron will be physically represented by different bonding patterns: edges between two hexagons will be double carbon bonds; those between a pentagon and a hexagon will be single bonds. Ab initio calculations for C_{60} by de Coulon, Martins, and Reuse (1992) confirm this physical geometry and predict average bond lengths of 1.386 Å and 1.453 Å, respectively, for the double and single bonds. These values, typical of aromatic carbon-carbon bonds, are in good agreement with the bond lengths deduced by Yannoni et al. (1991) from nuclear magnetic resonance (1.40 ± 0.015 Å and 1.45 ± 0.015 Å). Hawkins et al. (1991), using x-ray diffraction, have obtained bond lengths of 1.388(5) Å and 1.432(9) Å. These values for the single and double bond lengths and the molecular geometry lead to an average diameter of 7.1 Å for C_{60}.

Figure 2.2. C_{60}: A carbon atom occupies each vertex of the regular truncated icosahedron.

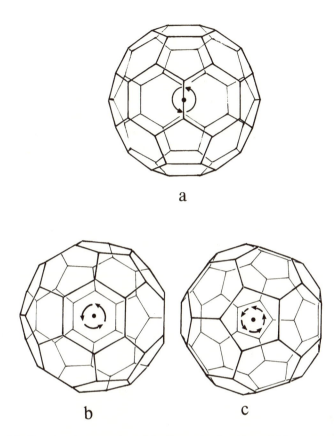

Figure 2.3. Rotational symmetries of the C_{60} molecule. (a) twofold axes pass through opposing pairs of hexagon-hexagon edges; (b) threefold axes connect midpoints of hexagonal faces related by inversion; (c) fivefold axes connect midpoints of pentagonal faces related by inversion. *Source*: Chung and Sternberg (1995).

The high symmetry of C_{60} means that symmetry considerations play a very important role in the understanding of the electronic and vibrational properties of the molecule. In reviewing symmetry properties of C_{60}, it is natural to begin with the symmetry operations for icosahedral symmetry. This group of operations has sixty elements: fifty-nine rotations, in which *all* vertices of the C_{60} cage move, and an identity element, in which every vertex remains fixed. (There are no nontrivial symmetry operations in which one or more vertices remain fixed while the others rotate.) Rotational symmetries of C_{60} are defined in terms of three sets of axes: fifteen axes of twofold symmetry, ten axes of threefold symmetry, and six axes of fivefold symmetry (Fig. 2.3).

The twofold axes are most easily defined in terms of the carbon double bonds (hexagon-hexagon edges). Every double bond is matched by another double bond on the opposite side of the molecule. That is, each double bond is mapped to another double bond by inversion through the molecule's center (Fig. 2.3 [a]). Each such pair of bonds is associated with a twofold axis of rotation, which passes through the midpoints of both bonds. There are thirty double bonds, resulting in fifteen twofold axes of rotation, for a total of fifteen twofold rotational symmetry operations. The threefold symmetry axes lie along lines that connect the midpoints of hexagonal faces paired by inversion (Fig. 2.3 [b]). The truncated icosahedron possesses twenty hexagonal faces, and thus there are ten pairs of such faces for ten threefold axes. These axes are threefold symmetric rather than sixfold, due to the alternation of single and double carbon bonds along the edges of each hexagonal face. Rotation by $120°$ and $240°$ for each axis produces twenty unique threefold rotational symmetry operations. Finally, the six fivefold axes lie along lines that connect the midpoints of opposing pentagonal faces on the molecule (Fig. 2.3 [c]). Each pentagonal face in a pair is related to its partner by inversion, and any rotation by a multiple of $72°$ about one of the fivefold axes maps the C_{60} molecule into itself. There are thus twenty-four unique fivefold rotational symmetry operations. The fifty-nine rotational symmetry operations $(15 + 20 + 24)$ and the identity operation define the sixty elements of the icosahedral group, I. We must include one more symmetry operation, however, before defining the full symmetry group for C_{60}.

As we have just seen, the C_{60} molecule possesses inversion through its center as a further class of symmetry operations. Interchanging any of the sixty carbon atoms with its opposite by inversion leaves C_{60} unchanged. Thus we can include P, the parity operator, as an additional symmetry operator for C_{60}. Combining P with the 60 elements of I creates a new 120-element group I_h, the full icosahedral group. It is this group that is the symmetry group for the C_{60} molecule.

The group I_h, as indicated by its name, is also the group of symmetry operations of the regular icosahedron shown in Figure 2.1(a), as it is also the symmetry group of the regular dodecahedron. The dodecahedron has twelve pentagonal faces; a fivefold axis passes through the center of each face; a threefold axis passes through each vertex, where three faces meet; and a twofold axis passes through the center of each edge.

2.3 IRREDUCIBLE REPRESENTATIONS OF I_h

Knowledge of the vibrational modes or electronic orbitals of C_{60} requires further information about the full icosahedral group, I_h. We shall make no pretence here of giving an introduction to group theoretical principles, but where group theoretical methods are to be used, they will be described as fully as possible.

TABLE 2.1

Character Table for I_h

In the table, $r = (1 + \sqrt{5})/2$; r is the golden ratio, satisfying the equation $r^2 = r + 1$.

I_h	E	$12C_5$	$12C_5^2$	$20C_3$	$15C_2$	P	$12S_{10}^3$	$12S_{10}$	$20S_3$	$15\sigma_v$
A_g	+1	+1	+1	+1	+1	+1	+1	+1	+1	+1
T_{1g}	+3	$+r$	$1-r$	0	−1	+3	r	$1-r$	0	−1
T_{2g}	+3	$1-r$	$+r$	0	−1	+3	$1-r$	r	0	−1
G_g	+4	−1	−1	+1	0	+4	−1	−1	+1	0
H_g	+5	0	0	−1	+1	+5	0	0	−1	+1
A_u	+1	+1	+1	+1	+1	−1	−1	−1	−1	−1
T_{1u}	+3	$+r$	$1-r$	0	−1	−3	$-r$	$r-1$	0	+1
T_{2u}	+3	$1-r$	$+r$	0	−1	−3	$r-1$	$-r$	0	+1
G_u	+4	−1	−1	+1	0	−4	+1	+1	−1	0
H_u	+5	0	0	−1	+1	−5	0	0	+1	−1

Source: Dresselhaus, Dresselhaus, and Eklund (1992).

We start by setting out the character table for this point group in Table 2.1. The leftmost column lists the ten irreducible representations of I_h. These are subscripted according to whether the basis states of their matrix representations are even (g) or odd (u) under inversion (when operated on by the parity operator P). The dimensionality of each irrep can be seen by the character of the identity class (E): A irreps are one-dimensional; T irreps are three-dimensional; there are two four-dimensional G irreps, and two five-dimensional H irreps. The various classes of group elements are listed horizontally along the top of Table 2.1: E (identity), $15C_2$ (all the elements for rotations about the fifteen twofold axes), $20C_3$ (all the rotations about the ten threefold axes), $12C_5$ and $12C_5^2$ (rotations about the six fivefold axes), and P (the inversion or parity element). The remaining classes involve elements that combine inversion and rotation. To be useful, the character table needs to be supplemented by sample sets of functions that transform among themselves and form bases for the various irreps. For instance, the components of a vector $\{x, y, z\}$ form a base for the irrep T_{1u}, as can be seen by noting that the traces of the rotation matrices for the operations in each class are identical to the characters listed for T_{1u}. The whole set of bases is most conveniently defined in terms of spherical harmonics, angular momentum wave functions. The classification of the functions in terms of the irreps of I_h is shown in Table 2.2. This table shows that not only do the components of a vector corresponding to $L = 1$ form a basis for T_{1u}, but that the set of five $L = 2$ states forms a basis for H_g. Beyond that, it is necessary to

TABLE 2.2

Spherical Harmonics $\{Y_{LM}\}$ reduced to
Irreps of I_h

The set of spherical harmonics $\{Y_{LM}\}$, $(2L + 1)$ in number (with L given), transform according to the irreps of I_h listed below.

L	Irreps of I_h
0	A_g
1	T_{1u}
2	H_g
3	$T_{2u} \oplus G_u$
4	$G_g \oplus H_g$
5	$T_{1u} \oplus T_{2u} \oplus H_u$
6	$A_g \oplus T_{1g} \oplus G_g \oplus H_g$
7	$T_{1u} \oplus T_{2u} \oplus G_u \oplus H_u$
8	$T_{2g} \oplus G_g \oplus 2H_g$
9	$T_{1u} \oplus T_{2u} \oplus 2G_u \oplus H_u$
10	$A_g \oplus T_{1g} \oplus T_{2g} \oplus G_g \oplus 2H_g$
11	$2T_{1u} \oplus T_{2u} \oplus G_u \oplus 2H_u$
12	$A_g \oplus T_{1g} \oplus T_{2g} \oplus 2G_g \oplus 2H_g$
13	$T_{1u} \oplus 2T_{2u} \oplus 2G_u \oplus 2H_u$
14	$T_{1g} \oplus T_{2g} \oplus 2G_g \oplus 3H_g$
15	$A_u \oplus 2T_{1u} \oplus 2T_{2u} \oplus 2G_u \oplus 2H_u$

Source: Fowler and Woolwich (1986).

form the bases out of linear combinations of the angular momentum eigenstates, and this is done in the Appendix (E.1). The full icosahedral symmetry of the molecular complex makes it convenient to label the electronic and vibrational states of the complex by irreps of I_h. Electron-phonon interactions that couple electronic and vibrational coordinates will not change this: the Hamiltonian for such a system necessarily has full icosahedral symmetry (under simultaneous transformation of *both* electronic and vibrational operators), so that the exact eigenstates associated with a particular vibronic energy level must transform as basis functions for an irrep of I_h. In the following chapters, we will use irrep labels in classifying the various Jahn-Teller systems in icosahedral complexes. To distinguish the electronic and vibrational parts of a system, we will employ uppercase or lowercase letters: uppercase for electronic and lowercase for vibrational. Thus, $T_1 \otimes h$ represents the interaction of an electronic triplet state with five h-type vibrational modes—a case we will consider at length in Chapter 3.

TABLE 2.3

Half-Integral Angular Momentum
States Classified According to the
Irreps of I_h

J	irreps
1/2	Γ_6
3/2	Γ_8
5/2	Γ_9
7/2	$\Gamma_7 \oplus \Gamma_9$
9/2	$\Gamma_8 \oplus \Gamma_9$
11/2	$\Gamma_6 \oplus \Gamma_8 \oplus \Gamma_9$
13/2	$\Gamma_6 \oplus \Gamma_7 \oplus \Gamma_8 \oplus \Gamma_9$
15/2	$\Gamma_8 \oplus 2\Gamma_9$

Source: Judd (1957).

2.3.1 Spin Representations: the Double Group

In order to handle the transformations of electron spin eigenstates under the operations of a point group, it is usual to double the size of the group and introduce a new set of irreps that are related to the half-integral angular momentum states in the same way as the original irreps are related to the integral angular momentum states. Each element of the original group gives rise to two elements of the double group, and the relationship is the same as between the single element of the rotation group, SO(3), and the corresponding pair of elements of SU(2). These spin representations are listed by Judd (1957) along with the character table and the classification of half-integral angular momentum states in terms of these representations. There are four extra representations, of degeneracy 2, 2, 4, and 6, and we shall follow Judd in calling them Γ_i ($i = 6, 7, 8, 9$). This notation assorts a little oddly with the use of A, T_1, T_2, G, and H for Judd's Γ_i ($i = 1 \ldots 5$), but it has the advantage of distinguishing between states originating from integral and half-integral angular momenta. The classification of the half-integral angular momentum states under I_h is shown in Table 2.3. As the table indicates, bases for Γ_6, Γ_8 and Γ_9 can be constructed from complete sets of J states, but a basis for Γ_7 requires a subset. The character table for the double group is given in Appendix D (Table D.2).

2.4 ELECTRON-PHONON INTERACTIONS

In icosahedral complexes, where high symmetry creates high degrees of degeneracy in the electronic states, the Jahn-Teller electron-phonon interaction is inevitably active. Jahn-Teller interactions are typically termed *linear* or *quadra-*

TABLE 2.4

The Kronecker Products of Irreps of I_h
The Kronecker product for T holds for T_1 or T_2. The subscripts on H represent multiplicity labels to these irreps.

$T \otimes T = [A \oplus H]_S \oplus \{T\}_A$

$G \otimes G = [A \oplus G \oplus H]_S \oplus \{T_1 \oplus T_2\}_A$

$H \otimes H = [A \oplus G \oplus H_2 \oplus H_4]_S \oplus \{T_1 \oplus T_2 \oplus G\}_A$

$\Gamma_8 \otimes \Gamma_8 = \{A \oplus H\}_A \oplus [T_1 \oplus T_2 \oplus G]_S$

$\Gamma_9 \otimes \Gamma_9 = \{A \oplus G \oplus H_2 \oplus H_4\}_A \oplus [2T_1 \oplus 2T_2 \oplus G \oplus H]_S$

tric, depending on whether the normal mode coordinates enter the interaction in linear or quadratic fashion. It is sufficient for our purposes at this point to limit the discussion to linear Jahn-Teller interactions, since these are capable of illustrating the general features of vibronic coupling in icosahedral symmetry.

We begin by considering an electronic state within icosahedral symmetry. If the state is represented by an irrep Γ of I_h, then there exists a set of basis states of dimension $|\Gamma|$ that span the electronic space and that transform among themselves as basis vectors of the irrep Γ. (In a similar manner, $|p_x\rangle$, $|p_y\rangle$, and $|p_z\rangle$ span the electronic triplet state in the more familiar SO(3) symmetry.) We will denote the abstract representation of these states by the set $\{|u_i^\Gamma\rangle\}$, with $i = 1, 2, \ldots, |\Gamma|$. (In our notation, $|\Gamma|$ denotes the dimensionality of irrep Γ.)

The normal mode coordinates of the nuclear framework can be represented by a set of coordinates Q_λ^Λ ($\lambda = 1, 2, \ldots, |\Lambda|$), which transform among themselves under the icosahedral symmetry operations as the irrep Λ of I_h. Symmetry considerations will allow these coordinates to couple to the electronic Γ state if they are of even parity and if the symmetric Kronecker square $[\Gamma \otimes \Gamma]_S$ contains Λ in its decomposition. Normal modes that meet these conditions are said to be Jahn-Teller active. For the irreps deriving from half-integral spin states, on the other hand, the Jahn-Teller-active modes are those contained in the antisymmetric Kronecker square, though still of even parity. These states are always doubled by time-reversal or Kramers degeneracy, which cannot be split by a Jahn-Teller interaction, so in I_h only Γ_8 and Γ_9 are candidates for a Jahn-Teller splitting. Herzberg (1966) gives the Kronecker products of the nontrivial irreps of I_h as are shown in Table 2.4. The symmetric parts of these products show, for example, that h_g vibrational modes will be the only Jahn-Teller-active modes able to couple to T electronic states. We can also see that electronic quartet (G) states will be able to couple to both g_g and h_g normal modes.

The five-dimensional irrep H occurs twice in the symmetric part of the Kronecker product H⊗H and in the antisymmetric part of $\Gamma_9 \otimes \Gamma_9$ in Table 2.4, where the two occurences are distinguished with different subscripts as H_2 and H_4. This complication means that the mode of coupling of an h_g mode is not fully defined by its symmetry label. Where this occurs, the ambiguity is covered by treating every mode as a linear combination of two modes that are differently coupled, h_2 and h_4.

Returning to a more general formulation, let us now consider a general linear interaction between electronic and vibrational degrees of freedom,

$$H^{JT} = \sum_{\Lambda,\lambda} V_\lambda^\Lambda Q_\lambda^\Lambda. \tag{2.1}$$

The Q_λ^Λ are the Jahn-Teller-active normal mode coordinates, and the V_λ^Λ are electronic operators. Both transform according to the irreps Λ of I_h, and this form is chosen so that H^{JT} is an icosahedral scalar, as it has to be if the Hamiltonian is to be invariant under the operations of I_h. The subscript λ ($\lambda = 1, 2, \ldots, |\Lambda|$) indexes a set of basis functions for Λ, and the indexing is kept consistent between these different realizations of the irrep. To obtain a matrix representation of (2.1), we place the interaction between the electronic states for the complex in its high- symmetry (icosahedral) configuration: $|u_i^\Gamma\rangle$, with $i = 1, 2, \ldots, |\Gamma|$. Within these states, the interaction matrix is

$$H_{ij}^{JT} = \sum_{\Lambda,\lambda} Q_\lambda^\Lambda \langle u_i^\Gamma | V_\lambda^\Lambda | u_j^\Gamma \rangle. \tag{2.2}$$

We now apply the Wigner-Eckart theorem to (2.2) in order to rewrite the interaction Hamiltonian in terms of a matrix equation,

$$\mathbf{H}^{JT} = \sum_{\Lambda,\lambda} \langle u^\Gamma || V^\Lambda || u^\Gamma \rangle \, Q_\lambda^\Lambda \, \mathbf{U}^\Gamma(\Lambda\lambda). \tag{2.3}$$

In (2.3), the reduced matrix element, $\langle u^\Gamma || V^\Lambda || u^\Gamma \rangle$, measures the strength of the electron-phonon coupling and is independent of the particular choice of electronic basis states. It is the symmetric matrices $\mathbf{U}^\Gamma(\Lambda\lambda)$ that contain the details of the coupling between the electronic states (Γ) and vibrational modes (Λ). Methods of constructing the $\mathbf{U}^\Gamma(\Lambda\lambda)$ can be found in Golding (1973) and in Cullerne, Angelova, and O'Brien (1995). These matrices are given in Appendix E (E.1.3) in a form that is dictated by the labelling of the bases, also given there, and this is the choice used in this book. Equation 2.3 and Table 2.4 tell us that a T electronic state will couple to the Jahn-Teller-active h modes through the five 3×3 matrices $\mathbf{U}^T(h,\alpha)$ ($\alpha = 1, \ldots, 5$). (Here, "h" is the five-dimensional irrep for the phonon space.) Similarly, the symmetric part of G⊗G shows that G electronic states will couple to g vibrational modes via four 4×4 matrices, $\mathbf{U}^G(g, \beta)$ ($\beta = 1, \ldots, 4$), and to h modes through the five 4×4 $\mathbf{U}^G(h, \gamma)$ ($\gamma = 1, \ldots, 5$). Interactions with electronic quintet states labelled by H are more complicated due to the Kronecker multiplicity mentioned earlier.

To evaluate these interactions, we must distinguish between the two realizations of the H irrep appearing in $[H \otimes H]_S$. In doing this, it is useful to consider the parentage of H with representations of SO(3). These are shown in Table 2.2. As the $L = 2$ entry shows, the five spherical harmonics for $L = 2$ transform as a realization of H, which we shall denote by H_2. A second realization can be obtained using the $L = 4$ spherical harmonics, and we will distinguish it by H_4. The electronic operators $V_{\lambda}^{H_2}$ and $V_{\lambda}^{H_4}$ must then also be written in terms of operators transforming as spherical harmonics (spherical tensor operators) in a similar way. The action of these operators on the electronic states is different, as required, and is shown in Appendix E (E.16 and E.17). This is an arbitrary choice, but it is convenient to use, and we shall usually hold to it in this book.

2.5 VIBRATIONAL MODES AND THEIR SYMMETRIES

As we have seen, normal modes can be classified by the irreps of I_h, since the modes are excitations in a Hamiltonian system that is invariant under all the symmetry transformations of the group. Section 2.4 has shown which normal modes are able to couple to which electronic states through a linear Jahn-Teller interaction. Under the influence of a Jahn-Teller interaction, an icosahedral molecule will distort, or be disposed to distort, its molecular configuration. Appropriate distortions must have the point symmetry of a subgroup of I_h and, in addition, must be the result of a Jahn-Teller-active normal mode. The distortions that are most likely to appear as stationary points on a potential energy surface are shown in Figure 2.4 with the normal modes that generate them. As indicated, the fourfold modes g_g are capable of reducing I_h symmetry to either T_h or D_{3d}, and the fivefold h_g modes can produce distortions of D_{2h}, D_{3d}, or D_{5d} symmetry. The modes are both even (*gerade*) and the subgroups must possess inversion symmetry, since I_h contains P, the parity operator. Thus D_{2h} is an allowed distortion symmetry, but D_{2d} is not. As we shall see, stable distortions of both D_{5d} and D_{3d} symmetry do appear under linear Jahn-Teller coupling, but a D_{2h} distortion can only be stabilized by high order perturbation processes or by external fields. In addition, external fields can on occasion stabilize a distortion of symmetry as low as C_i. A full discussion of this topic and the group theory involved is given by Ceulemans and Vanquickenbourne (1989).

In a large molecule such as C_{60}, there will be several modes belonging to each irrep, and the excitation of each one will result in a different and complex distortion of the molecule. In an undistorted icosahedral molecule, all the atoms lie on a spherical surface. It is convenient to think of each type of mode as producing a distortion of the spherical surface that carries the atoms with it and is proportional to the magnitude of a spherical harmonic at each point. This

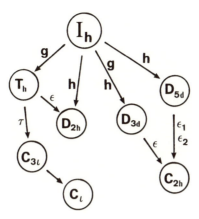

Figure 2.4. Reduction of I_h symmetry by vibronically stabilized distortions. The arrows point to the reduced symmetry group, and the arrows are labelled by the irrep of the normal mode that participates in the vibronic coupling. All normal modes are of even symmetry (e.g., h_g and τ_g). *Source*: Englman (1972).

provides a simple way of visualising the distortions. The appropriate spherical harmonic Y_{LM} is drawn from the list in Table 2.2. To get a simple picture with the fewest nodes, we can pick the smallest value of L for the symmetry in question. As Table 2.2 shows, h_g–type distortions can be represented by the $L = 2$ harmonics, and part of the $L = 4$ harmonics will contain g_g-type distortions. Using this method, we represent only vibrations in which the movement is perpendicular to the spherical surface. For instance, in Figure 2.5 we show an icosahedral cage under the action of an h_g distortion that results in D_{5d} symmetry. This is particularly simple to interpret, as the distortion, $3z^2 - r^2$, has an axis of symmetry along the z direction that is placed vertically in the picture. The form of the representation is to replace each vertex by either a cube or a sphere—a cube where the displacement is outwards and a sphere where the displacement is inwards—with the size of the cube or sphere being proportional to the size of the distortion. We have chosen this way of representing the distortion as being easier to see and interpret than a more realistic picture. Figure 2.6 shows a T_h distortion on a C_{60} cage, produced by g-type modes, in a similar representation. In Chapter 3 we shall be discussing systems in which the distortions move over the spherical surface of the molecule in the phenomenom called pseudo-rotation. This occurs with h-type modes, in which the axis of a $3z^2 - r^2$ distortion can take any orientation with respect to the symmetry axes of the molecule in the course of this rotation. As it moves, the actual symmetry of the distorted molecule changes even though the type of distortion of the underlying spherical surface remains unchanged. In

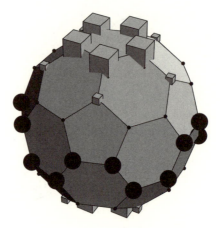

Figure 2.5. A D_{5d} distortion of a C_{60} cage produced by an h mode of $(3z^2 - r^2)$ type with its axis along a fivefold axis. Cubes represent displacements outwards and spheres inwards. The magnitude of the displacement is proportional to the cube/sphere size.

Figures 2.7 and 2.8 we show the D_{3d} and D_{2h} distortions that result from such a pseudo-rotation of the distortion shown in Figure 2.5 on a C_{60} cage.

Figure 2.9 by contrast shows the different D_{2h} distortion of a C_{60} produced by h-type modes at one of the type III saddle points of the $G \otimes h$ system described in Section 4.4.

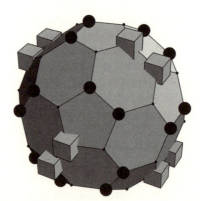

Figure 2.6. A T_h distortion of a C_{60} cage produced by g-type modes. Cubes represent displacements outwards and spheres inwards, with the displacement magnitude proportional to the cube/sphere size.

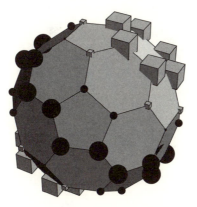

Figure 2.7. A D_{3d} distortion of a C_{60} cage produced by an h mode of $(3z^2 - r^2)$ type with its axis along a threefold axis. Cubes represent displacements outwards and spheres inwards, with the displacement magnitude proportional to cube/sphere size.

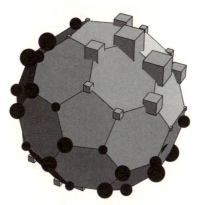

Figure 2.8. A D_{2h} distortion of a C_{60} cage produced by an *h* mode of $(3z^2 - r^2)$ type with its axis along a twofold axis.

2.6 HAM REDUCTION FACTORS

The concept of a Ham factor was introduced in Chapter 1, Section 1.3, in terms of the overlap of vibronic wave functions in two different minima. The lowest eigenfunctions of the $E \otimes \beta$ Hamiltonian (1.1) can be written

$$\Psi_1 = u_1 \times \psi_0(Q + k) \qquad \text{and} \qquad \Psi_2 = u_2 \times \psi_0(Q - k), \qquad (2.4)$$

where u_1 and u_2 are the two components of the doubly degenerate E electronic state, and ψ_0 is the lowest harmonic oscillator eigenfunction. Now if there is an operator, O, that acts between the electronic states, then the matrix elements

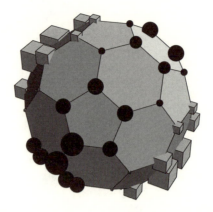

Figure 2.9. A D_{2h} distortion of a C_{60} cage produced by h modes at a type III saddle point of $G \otimes h$. (See the caption to Figure 2.7 for information on the symbols used.)

of O between the lowest pair of states after Jahn-Teller coupling will clearly be given by

$$\langle \Psi_1 | O | \Psi_2 \rangle = S \langle u_1 | O | u_2 \rangle, \quad \text{where} \quad S = \langle \psi_0(Q+k) | \psi_0(Q-k) \rangle. \quad (2.5)$$

The overlap of the displaced harmonic oscillator wave functions, S, which goes from 1 to 0 as k increases from 0 to infinity, acts to decrease the effect of O in the ground states, and so it is called a reduction factor. This simple example of a reduction factor is an example of a more general class whose use was first pointed out by Ham (1965). The generality derives from the Wigner-Eckart theorem, which we have already used in setting out the general form of the Jahn-Teller interaction in Section 2.4. The aspect of this theorem used here concerns the set of matrix elements of the form $\langle \phi_i^{\Gamma} | V_j^{\Lambda} | \phi_k^{\Gamma} \rangle$, where $\{\phi_i^{\Gamma}\}$ is a set of bases for the Γ irrep of the group and $\{V_j^{\Lambda}\}$ is a set of operators transforming as the Λ irrep of the group. The theorem states that these matrix elements are all related to each other by symmetry, so that there remains only one free parameter to be fixed by the actual choice of bases and operator, ϕ and V. Thus we can write

$$\langle \phi_i^{\Gamma} | V_j^{\Lambda} | \phi_k^{\Gamma} \rangle = \langle \phi^{\Gamma} || V^{\Lambda} || \phi^{\Gamma} \rangle \begin{pmatrix} \Gamma & \Lambda & \Gamma \\ i & j & k \end{pmatrix}, \quad (2.6)$$

where $\langle \phi^{\Gamma} || V^{\Lambda} || \phi^{\Gamma} \rangle$, the reduced matrix element, is the single free parameter, and the symbol to the right of it represents a set of numbers determined entirely by symmetry. When this theorem is applied to Jahn-Teller coupled states, the ϕ_i^{Γ} are the ground states, such as Ψ_1 and Ψ_2 above, and the V_j^{Λ} are a set of operators that operate in the original uncoupled electronic states, such as O above. As the strength of the Jahn-Teller coupling changes, the symmetry type of the ground states almost always remains the same, so the only part of (2.6) that changes is the reduced matrix element.[1] The Ham factor for a particular coupling strength is then defined as the ratio of the reduced matrix element to the reduced matrix element at zero coupling. This Ham factor would then be written $K(\Lambda)$. In the example from $E \otimes \beta$, the operator O might be an external stress tending to produce a B_{2g} distortion as shown in Figure 1.2, which has the pattern of matrix elements described for O. The reduction factor S is thus the Ham factor $K(B_{2g})$. This calculation also demonstrates the phenomen of quenching; the effect of this particular stress is reduced by the Jahn-Teller coupling, and may be totally quenched if the coupling is strong. The usefulness of calculating a set of Ham factors for any particular Jahn-Teller system lies in the link they form between experiment and theory. In principle, the existence of Jahn-Teller coupling could be investigated by measuring the effect of external probes, such as stress, and comparing the results with the calculated Ham factor. The number of different Ham factors sets an upper limit on the amount of information that could be obtained in this way. It is also useful to know which operators will be quenched by the Jahn-Teller interaction and which will not.

2.7 ICOSAHEDRAL JAHN-TELLER SYSTEMS

We have already alluded to some of the possible Jahn-Teller interactions in icosahedral symmetry. Here we will briefly list some of the possible linear interactions, with special regard for the symmetries of the various interaction Hamiltonians. (All must be scaler under icosahedral symmetry operations, but this does not preclude their possessing symmetries of higher order.) Early work on the symmetries of several of these systems was carried out by Khlopin, Polinger and Bersuker (1978) and by Pooler (1980). These systems are all discussed in detail in subsequent chapters.

2.7.1 $T \otimes h$ and $p^n \otimes h$

This interaction, discussed in Chapter 3, describes a triplet electronic state coupled to a set of five nuclear vibrations that transform as the H irrep. The Hamiltonian has rotational symmetry in a three-dimensional subspace of the five-dimensional phase space, and the low-energy states are rather closely spaced, being pseudo-rotational. A mathematically equivalent form of this system has been well studied: the $T \otimes (\epsilon \oplus \tau_2)_{eq}$ system in cubic symmetry—an electronic triplet equally coupled to a set of doubly-degenerate ϵ vibrational modes and a set of triply degenerate τ_2 vibrational modes. Al_{13} and the C_{60} anion C_{60}^{-} provide possible realizations of this interaction, as does C_{60}^{*}, since its LUMO

---

[1] Some exceptions arise in $H \otimes h$ and will be discussed in Chapter 5.

and LUMO+1 are triplet states. Multiple occupation of the T orbital leads to the system that we call $p^n \otimes h$, the model system for C_{60}^{-n}, with $n = 2, 3, 4$.

2.7.2 $G \otimes (g \oplus h)$

Because the symmetric square of G contains both G and H, the general Jahn-Teller system includes coupling to vibrational modes of both these types. Pooler (1980) showed that this Hamiltonian is invariant under SO(4) for equal couplings to both modes, with the result that the system possesses an energy minimum along a four-dimensional continuum of the phase space. Working from the equal-coupling regime, Ceulemans and Fowler (1989) analyzed the geometry of the APES's of this system, and Cullerne and O'Brien (1994) have analyzed its ground state properties at strong coupling (Section 4.5). For $G \otimes g$ (Section 4.2) the lowest APES has minima as well as threefold degeneracies at points corresponding to distortions of tetrahedral symmetry, and there are saddle points and degeneracies at distortions of D_{3d} symmetry. For $G \otimes h$ (Section 4.4) the lowest APES has ten points of minimum energy, corresponding to distortions of D_{3d} symmetry. These minima lie at the vertices of a dodecahedron embedded in the five-dimensional phase space defined by the normal coordinates, h_i ($i = 1, \ldots, 5$). Degeneracies also occur at points corresponding to D_{5d} distortion, and saddle points lie at distortions of D_{2h} symmetry. The saddle points nearest in energy to the minima lie near the centers of the edges of the dodecahedron. Ceulemans and Fowler have suggested that the $G \otimes (g \oplus h)$ interaction—or the more restricted $G \otimes g$ or $G \otimes h$ interactions—might occur in the fullerene cations C_{20}^{+} and C_{80}^{+}. The atomic cluster Si_{13}^{+} might also be a physical setting for an electronic-quartet-based Jahn-Teller interaction.

2.7.3 $H \otimes (g \oplus h)$

As discussed in Section 2.4, the added complication of multiplicity in the $H \otimes H$ Kronecker product means that the coupling of a quintet vibrational mode must involve a sum of interaction matrices with two different coupling constants in order to represent general coupling to an electronic quintet. For $H \otimes g$ (Section 5.5), the lowest APES possesses ten energetically-equivalent minima at points in phase space corresponding to D_{3d} distortions of the icosahedron. Other points on the lowest APES mark out other D_{3d} distortions for which degeneracies exist between the lowest APES and higher surfaces. (This second set of D_{3d} distortions are not energy minina.) For $H \otimes h$ (Section 5.3), the points of minimum energy on the lowest APES are either at the six D_{5d} distortions or the ten D_{3d} distortions, according to the choice of interaction matrix. In an exceptional case, $H \otimes h_2$ (Section 5.4), the symmetry for the Hamiltonian is SO(3), and the lowest APES is a four-dimensional hypersphere in the five-dimensional phase space. This APES is degenerate with the APES next highest in energy over a three-dimensional subspace. This degree of degeneracy leads to closely-spaced low-lying energy states. Other possibilities, $H \otimes (g \oplus h_4)_{eq}$ (Section 5.6) and $H \otimes (g + h_2 + h_4)_{eq}$ (Section 5.7), have Hamiltonians invariant under SO(3) and SO(5) symmetry respectively (Pooler 1980). In both cases the lowest APES has SO(5) symmetry. Possible physical examples of $H \otimes (g \oplus h)$ include the Si_{12} cluster and mono-cation of C_{60} (C_{60}^{+}).

3

T ⊗ *h* and p^n ⊗ *h*

3.1 INTRODUCTION

The two triplet irreducible representations T_1 and T_2 have the simplest type of Jahn-Teller interaction to be found in icosahedral symmetry. They are important in practice because the lowest empty molecular orbitals in the C_{60} molecule are T_{1u} triplets, so that in the negatively charged ion these states will contain unpaired electrons. Electronic triplets also occur in neutral C_{60}, where the lowest excited state is probably $^3T_{2g}$ and the first excited singlet state is probably $^1T_{2g}$. The T_1 and T_2 irreps are so similar in terms of their Jahn-Teller properties that it will often be unnecessary to distinguish between them in the rest of this chapter. There are, however, differences in the detail of the notation. For consistency of notation, what follows will be written for T_1, and we shall point out how the discussion applies to T_2. The adaptation is usually only a question of labelling; however, one important difference is that T_1 can have nonzero angular momentum whereas T_2 cannot. The symmetric square of either T representation with itself contains only the representations A and H, so only *h*-type vibrations need be considered. At this stage we assume that it is a good approximation to look at a Hamiltonian that operates entirely within this threefold degenerate electronic state coupled to a set of just five normal modes of vibration that span an H irreducible representation. A large molecule such as C_{60} has many sets of vibrations of H symmetry at different frequencies, so the use of a single set requires justification. This will be discussed in Chapter 6, and it suffices to say now that the use of a single effective mode can be well justified for studying ground state properties and for optical bands in solids. The single mode approximation is of much less use for analyzing the vibronic structure of spectra when individual lines are resolved. Theoretical work on aspects of the electronic triplet systems has been carried out by various researchers, and we shall cite their work at appropriate points in this chapter. All the work on T ⊗ *h* rests heavily on work that was originally done for a system in cubic symmetry, the system T ⊗ ($\epsilon \oplus \tau$)$_{eq}$ in which an electronic triplet is coupled simultaneously to two- and three-dimensional vibrational modes, with a particular set of parameters that produce an accidentally high symmetry. The Hamiltonian for this system is exactly the same as (3.1) below. It is only because molecules and centers with cubic symmetry were known and studied before icosahedral molecules that the work was done in the one context and not the other. Nonetheless, in the rest of this chapter the presentation will be in terms of the icosahedral group.

3.2 THE POTENTIAL ENERGY SURFACES

The Hamiltonian for $T_1 \otimes h$ can be written as

$$\mathcal{H} = -\frac{1}{2} \sum_{i=1}^{5} \frac{\partial^2}{\partial Q_i^2} + \frac{1}{2} \sum_{i=1}^{5} Q_i^2 + M^T(h). \qquad (3.1)$$

The first two terms represent the kinetic and potential energies of five simple harmonic oscillators, which together transform as the H irrep, with coordinates $Q_1, \ldots, Q_5$. We use reduced coordinates here in which $\hbar\omega = 1$. The last part represents the Jahn-Teller coupling between the vibrational modes and the set of three degenerate electronic states. This interaction takes the form of a 3×3 matrix, whose detailed structure can be predicted by symmetry, as described in Chapter 2, Section 2.4 (see also Appendix E). This Jahn-Teller interaction matrix is also particularly easy to find by using p state and d state wave functions for the electronic and vibrational bases (E.1, E.2). Then

$$M^T(h) = +\frac{1}{2}k \begin{bmatrix} Q_1 - \sqrt{3}Q_4 & -\sqrt{3}Q_3 & -\sqrt{3}Q_2 \\ -\sqrt{3}Q_3 & Q_1 + \sqrt{3}Q_4 & -\sqrt{3}Q_5 \\ -\sqrt{3}Q_2 & -\sqrt{3}Q_5 & -2Q_1 \end{bmatrix}, \qquad (3.2)$$

where k parametrizes the strength of the Jahn-Teller coupling, which is strong or weak according as $k \gg 1$ or $k \ll 1$. The three roots of this matrix are three energies. As we scan the five-dimensional $\{Q_i\}$ phase space, these three energies expand into three adiabatic potential energy surfaces (APESs), and if the potential energy of the original oscillators ($\frac{1}{2} \sum Q_i^2$), necessarily positive, is added, a minimum energy appears on the lowest APES. The depth of this energy minimum below the minimum energy of the uncoupled state is called the Jahn-Teller energy. It is worth noting at this point that the kinetic energy of the oscillators has not been included, and it will alter the actual energy of the ground state, which will always lie above the Jahn-Teller minimum. Even so, a good deal of information can be obtained before the kinetic energy is added, and we shall neglect it for the time being.

3.2.1 Rotational Symmetry of $T_1 \otimes h$

The interesting feature of the Hamiltonian (3.1) is that the minimum on the lowest APES does not just occur at isolated points in the $\{Q_i\}$ space but over a continuum that spans a three-dimensional spherical surface that is a subspace of the actual five-dimensional space. There are various ways of seeing this, but the most direct way is to make the substitution

$$Q_1 = Q\frac{1}{2}(3\cos^2\theta - 1), \tag{3.3}$$

$$Q_2 = Q\frac{1}{2}\sqrt{3}\sin 2\theta \cos\phi,$$

$$Q_3 = Q\frac{1}{2}\sqrt{3}\sin^2\theta \sin 2\phi,$$

$$Q_4 = Q\frac{1}{2}\sqrt{3}\sin^2\theta \cos 2\phi,$$

$$Q_5 = Q\frac{1}{2}\sqrt{3}\sin 2\theta \sin\phi,$$

which is a parametrization based on the $L = 2$ or d-state functions from which the matrix (3.2) was originally constructed. With this substitution we find that the matrix in (3.2) has the same eigenvalues, $-kQ$, $+kQ/2$, and $+kQ/2$, regardless of the values of θ and ϕ. The potential energy on the lowest APES can thus be written

$$V = -kQ + \frac{1}{2}Q^2, \tag{3.4}$$

which clearly has a minimum at $Q = k$ with a Jahn-Teller energy of $(1/2)k^2$. It is useful to interpret θ and ϕ as spherical polar angles so as to map this minimum onto a spherical surface. This is not a simple one-to-one mapping, as points on the spherical surface related by inversion correspond to the same set of $\{Q_i\}$. This feature of the mapping has important results. This is an example of a Jahn-Teller Hamiltonian that has an accidentally higher symmetry than is required by the symmetry of the original system, in this case SO(3). The SO(3) invariance of the Hamiltonian contains symmetry elements that are not present in the icosahedral group. In fact we have not yet proved that the Hamiltonian as a whole obeys this invariance. Pooler (1980) gives a general proof, but in what follows we shall do something more specific to this particular system. There is a parametrization of the $\{Q_i\}$ as in (3.3), but with two more angles added to make up the full five degrees of freedom. The derivation of this parametrization is described in Appendix F (F.1), and it is given by the following equations:

$$Q_1 = Q\left(\frac{1}{2}(3\cos^2\theta - 1)\cos\alpha + \frac{\sqrt{3}}{2}\sin^2\theta \sin\alpha \cos 2\gamma\right), \tag{3.5}$$

$$Q_2 = Q \left(\frac{\sqrt{3}}{2} \sin 2\theta \cos \phi \cos \alpha - \frac{1}{2} \sin 2\theta \cos \phi \sin \alpha \cos 2\gamma \right.$$

$$\left. + \sin \theta \sin \phi \sin \alpha \sin 2\gamma \right),$$

$$Q_3 = Q \left(\frac{\sqrt{3}}{2} \sin^2 \theta \sin 2\phi \cos \alpha + \frac{1}{2}(1 + \cos^2 \theta) \sin 2\phi \sin \alpha \cos 2\gamma \right.$$

$$\left. + \cos \theta \cos 2\phi \sin \alpha \sin 2\gamma \right),$$

$$Q_4 = Q \left(\frac{\sqrt{3}}{2} \sin^2 \theta \cos 2\phi \cos \alpha + \frac{1}{2}(1 + \cos^2 \theta) \cos 2\phi \sin \alpha \cos 2\gamma \right.$$

$$\left. - \cos \theta \sin 2\phi \sin \alpha \sin 2\gamma \right),$$

$$Q_5 = Q \left(\frac{\sqrt{3}}{2} \sin 2\theta \sin \phi \cos \alpha - \frac{1}{2} \sin 2\theta \sin \phi \sin \alpha \cos 2\gamma \right.$$

$$\left. - \sin \theta \cos \phi \sin \alpha \sin 2\gamma \right),$$

where the conditions $0 \le Q < \infty, 0 \le \alpha < \pi/3, 0 \le \gamma < \pi, 0 \le \theta < \pi/2$, and $0 \le \phi < 2\pi$ ensure that all possible distortions in the five-dimensional phase space are covered without repetition. It is also useful to find polynomials of the Q's that are invariant under SO(3) symmetry. One is quadratic,

$$I_2 = Q_1^2 + Q_2^2 + Q_3^2 + Q_4^2 + Q_5^2, \tag{3.6}$$

and the other is cubic,

$$I_3 = Q_1 \left(Q_1^2 - 3Q_4^2 - 3Q_3^2 + \frac{3}{2}Q_2^2 + \frac{3}{2}Q_5^2 \right) \tag{3.7}$$

$$+ 3\sqrt{3} \left(Q_3 Q_2 Q_5 + \frac{1}{2}Q_4 Q_2^2 - \frac{1}{2}Q_4 Q_5^2 \right).$$

Under the change of variables (3.5), they become

$$I_2 = Q^2, \tag{3.8}$$
$$I_3 = Q^3 \cos 3\alpha. \tag{3.9}$$

With this parametrization we can show the SO(3) invariance of the entire Hamiltonian by writing the matrix part of (3.1) in the form

$$
kT^{-1}
\begin{bmatrix}
\frac{1}{2}Q(\cos\alpha - \sqrt{3}\sin\alpha) & 0 & 0 \\
0 & \frac{1}{2}Q(\cos\alpha + \sqrt{3}\sin\alpha) & 0 \\
0 & 0 & -Q\cos\alpha
\end{bmatrix}
T,
$$
(3.10)

where T is an orthogonal matrix given by

$$
T =
\begin{bmatrix}
\cos\gamma & \sin\gamma & 0 \\
-\sin\gamma & \cos\gamma & 0 \\
0 & 0 & 1
\end{bmatrix}
\begin{bmatrix}
\cos\theta & 0 & -\sin\theta \\
0 & 1 & 0 \\
\sin\theta & 0 & \cos\theta
\end{bmatrix}
\begin{bmatrix}
\cos\phi & \sin\phi & 0 \\
-\sin\phi & \cos\phi & 0 \\
0 & 0 & 1
\end{bmatrix}.
$$
(3.11)

This transformation of the electronic basis shows that the roots of the matrix do not depend on the values of three of the angular parameters, (γ, θ, ϕ). The matrix T is in the form of a rotation expressed in terms of the Euler angles (γ, θ, ϕ), so for every set of values of the $\{Q_i\}$, the transformation that diagonalizes the matrix is a rotation in the space of the electronic bases. T has no effect on the representation of the first two terms of (3.1) as they are diagonal in the electronic basis. This then shows that the Hamiltonian can always be reduced to a standard form in which the potential energy is a function of Q and α only by a rotation in the electronic basis space. The lowest root of the matrix is $-kQ\cos\alpha$, constant over all the rotations T, and lowest of all when $\alpha = 0$. Putting $\alpha = 0$ in the transformation (3.5) recovers the substitutions (3.3) that placed the minimum on its spherical subspace. The transformation also gives eigenvectors for the matrix (3.2), since they may be read off directly as the rows of T. The eigenvector for the lowest root is

$$
|u\rangle = \sin\theta\cos\phi|\xi\rangle + \sin\theta\sin\phi|\eta\rangle + \cos\theta|\zeta\rangle,
$$
(3.12)

where $|\xi\rangle$, $|\eta\rangle$ and $|\zeta\rangle$ are the three components of the electronic T_1 base.

3.2.2 The Shape of the Distorted Molecule

The SO(3) invariance makes it very simple to describe the shape of the molecule or cluster in terms of a rotating distortion. At the point $\theta = 0$ on the spherical trough on the lowest APES, only Q_1 is nonzero, and the electronic basis is the z component of the equivalent p state. If the normal coordinates are realised as a set of quadrupole distortions of a sphere, as described in Chapter 2, then the Q_1 coordinate on its own corresponds to a spheroidal distortion with its axis along the $\theta = 0$ or z-axis. As the point on the spherical trough moves, so both the axis of the electronic p state and the principal axis of the spheroidal distortion follow the direction indicated by (θ, ϕ). This rotating distortion is commonly called a pseudo-rotation, and it is illustrated in Figure 3.1. This rotational symmetry

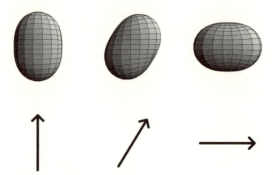

Figure 3.1. A pseudo-rotation of a distortion of a spherical surface. The principal axis of the distortion is shown with three different orientations, the axis of the associated electronic p states is shown below each distortion.

is easily broken by interactions of lower symmetry. We shall discuss terms in the molecular Hamiltonian, the warping terms that break the symmetry, in Section 3.2.3, but in addition there may be external influences. If the molecule is embedded in a crystal or in another matrix, there will be crystal fields as well as the fields of local strains. If such fields are not so large as to override the Jahn-Teller interaction, they will tend to alter the relative energies of the different distortions, so that one particular distortion is picked out as having the lowest energy. The measured properties of the molecule will then correspond to that particular distortion.

3.2.3 "Warping" of the Lowest APES

In the next section we shall go more carefully into the question of solutions to this Hamiltonian when the kinetic energy is included, but for the moment we should notice that having this accidently high symmetry in the Hamiltonian with linear coupling does not prevent other nonlinear terms being added to the Hamiltonian that will bring the symmetry down to icosahedral. One category of these terms can be put together by investigating higher-order polynomials in the $\{Q_i\}$ that are icosahedral invariants. This can be done using group-theoretical coupling coefficients. Physically these terms in the energy derive from inelastic terms in the interatomic forces. Because of the variety in the table of products of irreps in this group, there are many such invariants, which look very complicated when written out. However, as long as they arise only from perturbations that remain small compared with the Jahn-Teller energy itself, we need to evaluate them only on the trough at $\alpha = 0$. In addition to the invariant I_3 (3.7), there is another cubic polynomial, I_3', which is invariant

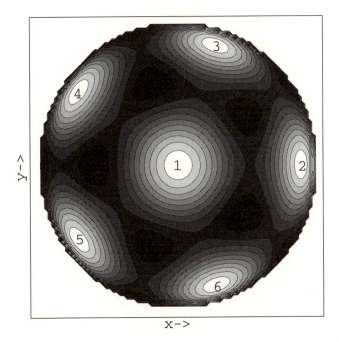

Figure 3.2. A contour plot of I'_3 or V_{icos} projected onto the sphere $\alpha = 0$. The six visible minima are numbered to define a numbering system for future use.

under icosahedral symmetry but not under SO(3). This polynomial is

$$I'_3 = Q_1 \left(Q_1^2 - \frac{3}{2} Q_4^2 + 3 Q_5^2 + 3 Q_2^2 - \frac{3}{2} Q_3^2 \right) \tag{3.13}$$

$$+ 3\sqrt{3} \left(Q_4 Q_3 Q_5 - \frac{1}{2} Q_4^2 Q_2 + \frac{1}{2} Q_3^2 Q_2 \right).$$

I'_3 is not constant over the $\alpha = 0$ surface, and is thus the lowest order polynomial to produce warping. It varies as shown in Figure 3.2, where the minima are numbered for later use. In this plot the sign chosen for I'_3 puts the minima on the spherical surface at the vertices of an icosahedron, the maxima at the vertices of a dodecahedron, and saddle points at the centers of the edges of either polyhedron. Changing the sign interchanges the maxima and minima, but leaves the saddle points where they were. Most of the various fourth order invariants produce a similar effect on this sphere. This introduction of a term that causes the energy to vary over a previously "flat" energy surface has been christened *warping*, and the energy variation produced has exactly the same angular variation as the lowest order homogeneous function of (x, y, z) that is of icosahedral symmetry. This function can be written

$$V_{icos} = 231z^6 - 315r^2z^4 + 105r^4z^2 - 5r^6 + 42z(x^5 - 10x^3y^2 + 5xy^4). \quad (3.14)$$

Another possible source of symmetry breaking is the quadratic Jahn-Teller effect. We have so far included in the Hamiltonian only terms in the electron-phonon interaction that are linear in the Q's, but this is really only the first term in a Taylor expansion of the vibronic interaction, and it makes sense to look at some higher order terms from the expansion.[1] This term in the Hamiltonian takes the form of a matrix similar to the existing linear Jahn-Teller matrix, $M^T(h)$, (3.2) but with entries that are quadratic in the $\{Q_i\}$. To construct this, it is necessary to find and use the sets of quadratic forms in the $\{Q_i\}$ that transform as the H irrep of the group—two sets because the symmetric product $H \otimes H$ contains H twice over. These matrices are given by Dunn and Bates (1995). Rather than using these matrices in their general form, we notice that we only need the expectation values on the trough at $\alpha = 0$, for which we already have the values of the $\{Q_i\}$ (3.3) and the electronic wave function (3.12). The cubic forms resulting from this calculation are exactly the same as the two considered in the previous paragraph, I_3 and I'_3. The conclusion to be drawn is that there exists one second-order Jahn-Teller effect that warps the lowest APES in a way that is indistinguishable from V_{icos}, and one, corresponding to I_3, that does not. If there is a large warping term, it is possible for the system to be trapped in one of the wells, so that it shows an effectively static Jahn-Teller distortion. This is what happens frequently in the better known case of E $\otimes$ ϵ coupling in cubic symmetry, which has an artificially high symmetry too (see Appendix C). In E $\otimes$ ϵ the symmetry is more often broken than not, and the warping terms must not be omitted from consideration. The effect of warping on the T $\otimes$ h eigenstates will be discussed later (Section 3.3.4).

3.2.4 Modification for $T_2 \otimes h$

The matrix for the $T_2 \otimes h$ interaction is given in Appendix E (E.12). It is related to (3.2) by the following substitutions:

$$\begin{aligned}
Q_1 &\to Q_1, \\
Q_2 &\to -Q_4, \\
Q_3 &\to -Q_5, \\
Q_4 &\to Q_2, \\
Q_5 &\to -Q_3.
\end{aligned} \quad (3.15)$$

Clearly the Hamiltonian can be made to look exactly the same as (3.10) by making a different choice of the assignment of the Q_i's in (3.5), after which

---

[1] The derivation of Jahn-Teller coupling from a Taylor expansion of the energy can be found in Englman (1972) or Bersuker and Polinger (1989).

all the argument about SO(3) invariance of the Hamiltonian follows through in the same way. The only difference is that we are now talking about a different spherical subspace of the five-dimensional phase space of the $\{Q_i\}$. It is interesting to note that whereas in $T_1 \otimes h$ the Jahn-Teller trough (the APES minimum) is given in terms of the invarients I_2 (3.6) and I_3 (3.7) by $I_2 = Q_0^2$, $I_3 = Q_0^3$ at a constant Q_0 with I_3' (3.13) doing the warping, in $T_2 \otimes h$ the trough is given by $I_2 = Q_0^2$, $I_3' = Q_0^3$ while I_3 does the warping.

3.3 THE GROUND STATES AT STRONG COUPLING

At strong coupling the best way to find details of the energies and other properties of the low-lying states is by using the full adiabatic approximation, including the kinetic energy (Appendix A). The change of phase of the electronic part of the wave function over the lowest APES is always a crucial piece of information, as it must be matched to the phase of the vibrational part to give a properly single-valued product wave function. The change of phase of the electronic part is related to Berry's geometric phase (Berry, 1984) described in Section 1.4, but for the states in this chapter, the results are more easily seen by direct inspection of the sign changes of wave functions without explicit reference to the Berry phase.

3.3.1 The Ground States in the Adiabatic Approximation

Using the results of the last two sections, we now apply the full adiabatic approximation (see Appendix A) to the Hamiltonian (3.1) by assuming a total wave function of the general form

$$\Psi = \psi(Q, \alpha, \gamma, \theta, \phi)u(Q, \alpha, \gamma, \theta, \phi, \mathbf{r}), \tag{3.16}$$

where $u(Q, \alpha, \gamma, \theta, \phi, \mathbf{r})$, is the electronic wave function in which the vibrational coordinates enter as parameters and $\mathbf{r}$ represents all the electronic coordinates. Here $u(Q, \alpha, \gamma, \theta, \phi, \mathbf{r})$ is just the eigenstate $|u\rangle$ that has already been calculated (3.12). When this Ψ is substituted into the Schrödinger equation for motion on the lowest APES, we get

$$-\frac{1}{2}[u\nabla^2\psi + 2\nabla\psi \cdot \nabla u + \psi\nabla^2 u] + \left(\frac{1}{2}Q^2 - kQ\cos\alpha\right)\psi u = E\psi u, \tag{3.17}$$

where $-\frac{1}{2}\nabla^2$ is the kinetic energy operator, the first term in (3.1). Applying closure to this equation with u gives

$$-\frac{1}{2}\nabla^2\psi - \nabla\psi \cdot \langle u|\nabla u\rangle - \frac{1}{2}\psi\langle u|\nabla^2 u\rangle + \left(\frac{1}{2}Q^2 - kQ\cos\alpha\right)\psi = E\psi. \tag{3.18}$$

Because $|u\rangle$ is real, $\langle u|\nabla u\rangle = 0$, but $\langle u|\nabla^2 u\rangle$ must be calculated and included. To use this, we need to have the kinetic energy operator in terms of these variables in (3.5), and this takes the form

$$
\mathcal{H}_{KE} = -\frac{1}{2}\left[Q^{-4}\frac{\partial}{\partial Q}\left(Q^4\frac{\partial}{\partial Q}\right) + \frac{1}{Q^2\sin 3\alpha}\frac{\partial}{\partial\alpha}\left(\sin 3\alpha\frac{\partial}{\partial\alpha}\right)\right]
$$
$$
+ \frac{1}{8Q^2}\left[\frac{\lambda_x^2}{\sin^2(\alpha - 2\pi/3)} + \frac{\lambda_y^2}{\sin^2(\alpha + 2\pi/3)} + \frac{\lambda_z^2}{\sin^2\alpha}\right], \quad (3.19)
$$

where $\{\lambda_x, \lambda_y, \lambda_z\}$ are the three components of an angular momentum operator λ within the phonon space. Explicitly,

$$
\lambda_x = i\cos\gamma\left(\cot\theta\frac{\partial}{\partial\gamma} - \csc\theta\frac{\partial}{\partial\phi}\right) + i\sin\gamma\frac{\partial}{\partial\theta},
$$
$$
\lambda_y = -i\sin\gamma\left(\cot\theta\frac{\partial}{\partial\gamma} - \csc\theta\frac{\partial}{\partial\phi}\right) + i\cos\gamma\frac{\partial}{\partial\theta}, \quad (3.20)
$$
$$
\lambda_z = i\frac{\partial}{\partial\gamma}.
$$

(This form of the kinetic energy is the same as that discussed by Bohr and Mottelson (1975) in connection with quadrupole oscillations in nuclei.) Next the kinetic energy operator (3.19) is adapted to operate in the neighborhood of the minimum surface by putting $\alpha = 0$ wherever this does not introduce infinities. We also make the substitution

$$
\psi \to \frac{\sqrt{\sin\alpha}}{Q\sqrt{\sin 3\alpha}}\psi, \quad (3.21)
$$

which puts this operator into a more familiar form for solving. The result of these two procedures is

$$
\mathcal{H}_{KE} = -\frac{1}{2}\left[\frac{1}{Q^2}\frac{\partial}{\partial Q}\left(Q^2\frac{\partial}{\partial Q}\right) + \frac{1}{Q^2\sin\alpha}\frac{\partial}{\partial\alpha}\left(\sin\alpha\frac{\partial}{\partial\alpha}\right)\right. \quad (3.22)
$$
$$
\left. + \frac{1}{Q^2\sin^2\alpha}\frac{\partial^2}{\partial\gamma^2} + \frac{f(\alpha)}{Q^2}\right]
$$
$$
- \frac{1}{6Q^2}\left[\frac{1}{\sin\theta}\frac{\partial}{\partial\theta}\left(\sin\theta\frac{\partial}{\partial\theta}\right) + \frac{1}{\sin^2\theta}\frac{\partial^2}{\partial\phi^2}\right],
$$

where $f(\alpha) = 2/3$ at $\alpha = 0$. This term comes from the substitution (3.21). To evaluate the term $-\frac{1}{2}\langle u|\nabla^2 u\rangle$ needed for equation (3.18), only the final line of (3.22) need be used because $|u\rangle$ depends only on θ and ϕ. This term is $\frac{1}{3Q^2}$, which exactly cancels the $-\frac{1}{2}\frac{f(\alpha)}{Q^2}$ term noted earlier. Thus the final form of the vibronic Schrödinger equation is

$$
-\frac{1}{2}\left[\frac{1}{Q^2}\frac{\partial}{\partial Q}\left(Q^2\frac{\partial}{\partial Q}\right)+\frac{1}{Q^2\sin\alpha}\frac{\partial}{\partial\alpha}\left(\sin\alpha\frac{\partial}{\partial\alpha}\right)\right. \tag{3.23}
$$
$$
\left.+\ \frac{1}{Q^2\sin^2\alpha}\frac{\partial^2}{\partial\gamma^2}-Q^2+2kQ\right]\psi
$$
$$
-\frac{1}{6Q^2}\left[\frac{1}{\sin\theta}\frac{\partial}{\partial\theta}\left(\sin\theta\frac{\partial}{\partial\theta}\right)+\frac{1}{\sin^2\theta}\frac{\partial^2}{\partial\phi^2}\right]\psi = E\psi.
$$

This equation is clearly separable: the first part is just the Hamiltonian of a three-dimensional harmonic oscillator with an origin displaced to $Q = k$, and the energy of its lowest state is $-\frac{1}{2}k^2+\frac{3}{2}$. The second part is the Hamiltonian of a rotator with moment of inertia $Q\sqrt{3}$, so the formula for the low-lying energies at strong coupling is

$$
E = -\frac{1}{2}k^2 + \frac{3}{2} + \left(\frac{1}{6k^2}\right)L(L+1), \tag{3.24}
$$

where L, the angular momentum quantum number, is a positive integer or zero. Finally we must look at the phase changes over the (θ,ϕ) surface. Inspection of the $\{Q_i\}$ (3.5) shows that they are repeated when

$$
\gamma \to \gamma + \pi \tag{3.25}
$$

and also under the inversion operation

$$
\theta \to \pi - \theta, \quad \phi \to \phi + \pi, \quad \gamma \to -\gamma. \tag{3.26}
$$

The electronic eigenstate (3.12) is seen to be invariant under (3.25) and to change sign under (3.26), so that the pseudo-rotational eigenstates must also change sign with the electronic eigenstate to preserve invariance. Thus invariance under these transformations requires L to be odd. As a result, the lowest energy level is a psuedo-rotational $L = 1$ state threefold degenerate, and higher states follow the sequence $L = 3, 5, 7, \ldots$. We should note that the lowest state is a T state, of the same symmetry as the electronic state we started with before the Jahn-Teller coupling was introduced. We shall see in Section 3.4 that a numerical method finds a ground state of the same symmetry at all coupling strengths.

3.3.2 The Ham Factors in T$_1$ ⊗ h

The Ham factors, defined in Section 2.6, are important because of their ability to link theory and experiment. In the T$_1$ ⊗ h system it is quite easy to find the

Ham factors in the strong coupling ground state, and in fact there is only one Ham factor that is of any interest. This is $K(H)$, which relates the effect of an operator of H symmetry on the vibronic ground states to the effect of the same operator in the original electronic states. This is relevant if we are looking at the effects of stress, since the set of stress operators will always transform like a d state, and in icosahedral symmetry this is the H representation. It is hard to see that any experiment could produce an exact value for $K(H)$, but the fact that it is nonzero, so that the ground state is split by stress, could be significant. The other possible Ham factor of interest, $K(T_1)$, gives the effect of the angular momentum operator in the ground state, and this is always 0 at strong coupling because the wave function is real. Looking at the square of T_1, we see that any operators with nonzero matrix elements must be of symmetry type A, T_1 or H. The Ham factor for an A-type operator is always 1. T_1 forms the antisymmetric square and so has to have a zero matrix element within the real electronic basis of the strongly coupled system as remarked above. Therefore only $K(H)$ lies between 0 and 1, and is in fact equal to 2/5 at strong coupling. The easiest way to find $K(H)$ is by picking on a Q_1 distortion as a typical H operator, represented by a matrix $\mathbf{U}^T(H, 1)$, which can be obtained from (3.1) by putting $Q_1 = 1$ and the rest of the $\{Q_i\} = 0$. It is

$$\mathbf{U}^T(H, 1) = \frac{1}{2}\begin{bmatrix} 1 & 0 & 0 \\ 0 & 1 & 0 \\ 0 & 0 & -2 \end{bmatrix}. \tag{3.27}$$

Thus the expection of $\mathbf{U}^T(H, 1)$ in the lowest state $|u\rangle$ (3.12) is

$$(\sin^2\theta - 2\cos^2\theta)/2, \tag{3.28}$$

and hence the expectation of $\mathbf{U}^T(H, 1)$ in a normalized $L = 1$, $M_L = 0$ state is

$$\langle\Psi_{1,0}|\mathbf{U}^T(H, 1)|\Psi_{1,0}\rangle = \frac{1}{2}\frac{\int\cos^2\theta(\sin^2\theta - 2\cos^2\theta)\sin\theta d\theta}{\int\cos^2\theta\sin\theta d\theta} \tag{3.29}$$

$$= -\frac{2}{5}.$$

The Ham factor is then given by

$$K(H) = \frac{\langle\Psi_{1,0}|\mathbf{U}^T(H, 1)|\Psi_{1,0}\rangle}{\langle\psi_{el,1,0}|\mathbf{U}^T(H, 1)|\psi_{el,1,0}\rangle} = \frac{2}{5}. \tag{3.30}$$

3.3.3 The Ham Factors in $T_2 \otimes h$

The square of the T_2 representation contains A, T_2, and H. The calculation of $K(H)$ proceeds exactly the same as in the previous paragraph, while T_2, which forms the antisymmetric part of the square, plays the same role as T_1 in $T_1 \otimes h$.

3.3.4 The Ground State with Warping

As so frequently happens, it is difficult to do a total analysis of the effect of warping, but it is quite easy to see what happens at the extremes of weak and strong warping fields, as will be set out here. To be specific: the warping field is weak or strong according to whether its energy is small or large compared with the splitting of the pseudo-angular momentum states. A more extended calculation than we give here is to be found in the work of Dunn and Bates (1995). These authors use a unitary transformation method to get expressions for energies at intermediate Jahn-Teller coupling and warping strengths, with the warping derived explicitly from the second order Jahn-Teller interaction. To start with the weak warping case, we introduce the field V_{icos} (3.14) as a perturbation into the pseudo-angular momentum states. The way angular momentum states split up under this perturbation is laid out in Table 2.2. For the states we are concerned with,[2] this list gives

$$
\begin{aligned}
L = & \\
1 \quad &\rightarrow \quad T_1 \\
3 \quad &\rightarrow \quad T_2 + G \\
5 \quad &\rightarrow \quad T_1 + T_2 + H \\
7 \quad &\rightarrow \quad T_1 + T_2 + G + H,
\end{aligned}
\tag{3.31}
$$

showing how the levels split up and, in particular, that the lowest state is unaltered to first-order in the warping strength. The Ham factors in the lowest state will thus also be unaltered to first-order. The actual arrangement of the split states will depend on the sign of V_{icos} which of course depends on the sign of the warping. If we started with the electronic state T_2, the decomposition of the pseudo-rotational states would not go exactly like angular momentum states, but this table would still give the correct assignment if the subscripts $_1$ and $_2$ were interchanged. To see what happens to the ground states with strong warping, we first note that the wave function will be strongly localised at points on the $\alpha = 0$ spherical surface that are either at the vertices of a dodecahedron or at the vertices of an icosahedron according to the sign of the warping potential as shown in Figure 3.2. The symmetry condition discussed in Section 3.3.1 still holds and requires the vibrational part of the wave function to change sign under inversion.

3.3.4.1 Warping and Tunneling with Icosahedral Wells

Most perturbation schemes predict that the distortions of D_{5d} symmetry give the lowest energy, so let us start by considering the wells at the icosahedral vertices. As before, the discussion is given in terms of a T_1 electronic state;

---

[2]These are the states with odd L because of the symmetry condition discussed in Section 3.3.1.

for a T_2 starting state we should interchange the subscripts $_1$ and $_2$ throughout. There are twelve wells, and so six linear combinations can be made that are odd under inversion. Let us use a notation that denotes a wave function that is strongly localised at the ith well as f_i, $i = 1, \ldots, 12$ and choose numbering such that well number $i \pm 7$ is diametrically opposite to number i. The odd states will then be linear combinations of $f_1 - f_7$, $f_2 - f_8$, and so on. The prediction of the symmetries of these linear combinations can be done in exactly the same way as one would predict the symmetries of normal modes or molecular orbitals on a set of atoms at the vertices of a icosahedron. In fact what we are looking for is the set of odd σ bonds, and these are T_1 and T_2. The appropriate linear combination for the two symmetries could be found by the use of projection operators, but here we shall use a method that will also indicate the relative energies of the two sets of states. These energies will depend on tunneling between the wells, which in turn will be influenced by the overlap, so we set up a matrix of overlaps between the $\{f_i\}$. This matrix can be written as

$$
S_I = \begin{bmatrix}
0 & S & S & S & S & S \\
S & 0 & S & -S & -S & S \\
S & S & 0 & S & -S & -S \\
S & -S & S & 0 & S & -S \\
S & -S & -S & S & 0 & S \\
S & S & -S & -S & S & 0
\end{bmatrix}, \tag{3.32}
$$

where the "bases" are $f_1 - f_7$, $f_2 - f_8$, $\ldots$, with the numbering as shown in Figure 3.2. An S has been entered for all neighbouring f_i's that have the same sign, and a $-S$ for those that have opposite signs. The eigenvalues of this matrix are $\sqrt{5}S$ three times and $-\sqrt{5}S$ three times, indicating that we can expect the sixfold degenerate ground state to split into two groups of three. (See Appendix B for a simpler version of this process.) These eigenvalues are not the actual energies. The differences in energy between the eigenstates in this situation will appear as the result of tunneling between the wave functions at the wells, as described in Appendix B. This tunneling is only important between wells that are nearest neighbours. It is always the case that a state of higher overlap will have lower tunneling energy, so we can identify the lowest triplet with the eigenvectors associated with the largest eigenvalue of S_I: $\sqrt{5}S$. The size of these energy differences produced by tunneling will decay exponentially with increasing warping energy. Because a set of bases for the T_1 representation is the set of components of a vector, (x, y, z), the linear combination of icosahedral minima that are also a basis for T_1 can be found by finding the values of x, y, and z at the minima. These are the positions of the vertices of an icosahedron, and given in terms of the angular parameters θ and

ϕ, as shown below.

	θ	ϕ		
1	0	any		
2	θ_I	0		
3	θ_I	$2\pi/5$	where	$\cos\theta_I = 1/\sqrt{5}.$ (3.33)
4	θ_I	$4\pi/5$		
5	θ_I	$6\pi/5$		
6	θ_I	$8\pi/5$		

Thus the unnormalised z component of the T_1 eigenvector is

$$(1, \cos\theta_I, \cos\theta_I, \cos\theta_I, \cos\theta_I, \cos\theta_I) \tag{3.34}$$

or, normalised,

$$\psi(T_1, z) = (\sqrt{5}, 1, 1, 1, 1, 1)/\sqrt{10}, \tag{3.35}$$

and it is easily verified that this corresponds to the eigenvalue $+\sqrt{5}S$ of (3.32), and hence to the ground triplet. The other two eigenvectors, produced in the same way, but using x and y instead of z, are

$$\sin\theta_I \times (0, 1, \cos(2\pi/5), \cos(4\pi/5), \cos(6\pi/5), \cos(8\pi/5))$$
$$\sin\theta_I \times (0, 0, \sin(2\pi/5), \sin(4\pi/5), \sin(6\pi/5), \sin(8\pi/5)). \tag{3.36}$$

Normalised these are

$$\psi(T_1, x) = (0, 4, \sqrt{5}-1, -\sqrt{5}-1, -\sqrt{5}-1, \sqrt{5}-1)/2\sqrt{10} \tag{3.37}$$

$$\psi(T_1, y) = (0, 0, 2, \sqrt{5}-1, -\sqrt{5}+1, -2)/2\sqrt{5-\sqrt{5}}.$$

These results, produced in a different way, are given by Wang et al. (1994) and by Dunn and Bates (1995) (but note that these latter authors make a different choice of axes). For the Ham factor, $K(H)$, we need to know the value of $\mathbf{U}^T(H, 1)$ at the minima. We use the fact that $\theta = 0$ at minimum no.1, and $\cos^2\theta = 1/5$ at all the other minima. We then find from (3.28) that at the minima

$$\mathbf{U}^T(H, 1) = (-5, 1, 1, 1, 1, 1)/5, \tag{3.38}$$

so its expectation in $\psi(T_1, z)$ is $-2/5$, leading to

$$K(H) = \frac{2}{5} \tag{3.39}$$

as before. The fact that $K(H)$ is the same for strong warping as for no warping at all does not necessarily mean that it is constant at intermediate strengths. In fact, numerical work shows it becomes somewhat smaller than 2/5. A similar situation occurs in $E \otimes \epsilon$; the Ham factor here becomes a little smaller at intermediate warping strength, and numerical calculations can be used to find a lower limit.

3.3.4.2 Warping and Tunneling with Dodecahedral Wells

In the case of the warping producing D_{3d} wells at the twenty vertices of a dodecahedron, there are ten states that are odd under inversion, and we can set up a matrix of overlaps in exactly the same way as for the icosahedral wells just discussed. Each vertex of a dodecahedron has three others as neighbours, so the overlap matrix has three nonzero entries in each row and column. The matrix can be written

$$
S_D = \begin{bmatrix}
0 & S & 0 & 0 & S & S & 0 & 0 & 0 & 0 \\
S & 0 & S & 0 & 0 & 0 & S & 0 & 0 & 0 \\
0 & S & 0 & S & 0 & 0 & 0 & S & 0 & 0 \\
0 & 0 & S & 0 & S & 0 & 0 & 0 & S & 0 \\
S & 0 & 0 & S & 0 & 0 & 0 & 0 & 0 & S \\
S & 0 & 0 & 0 & 0 & 0 & 0 & -S & -S & 0 \\
0 & S & 0 & 0 & 0 & 0 & 0 & 0 & -S & -S \\
0 & 0 & S & 0 & 0 & -S & 0 & 0 & 0 & -S \\
0 & 0 & 0 & S & 0 & -S & -S & 0 & 0 & 0 \\
0 & 0 & 0 & 0 & S & 0 & -S & -S & 0 & 0
\end{bmatrix}. \tag{3.40}
$$

The eigenvalues of this matrix are $\sqrt{5}S$ three times, 0 four times, and $-\sqrt{5}S$ three times: two triplets and a quartet, corresponding to states of T_1, G, and T_2 symmetry. We expect the T_1 state to be the lowest triplet, and set out to identify it. The values of θ and ϕ at these wells, which are at the vertices of a dodecahedron, are listed in Appendix G (G.4). The T_1 eigenvectors are found, as described for the icosahedral wells, by taking the values of x, y, and z at each well and normalising. They are

$$\psi(T_1, x) = (a_1, -a_2, -a_3, -a_2, a_1, a_5, -a_6, -a_7, -a_6, a_5), \tag{3.41}$$
$$\psi(T_1, y) = (a_8, a_9, 0, -a_9, -a_8, a_9, a_{10}, 0, -a_{10}, -a_9),$$
$$\psi(T_1, z) = (a_5, a_5, a_5, a_5, a_5, a_2, a_2, a_2, a_2, a_2),$$

where[3]

$$
\begin{array}{llll}
a_1 = 0.268999, & a_2 = 0.102749, & a_3 = 0.332502, & a_4 = 0.268999, \\
a_5 = 0.43525, & a_6 = 0.166251, & a_7 = 0.537999, & a_8 = 0.19544, \\
& a_9 = 0.316228, & a_{10} = 0.511667. &
\end{array}
$$

These are the eigenvectors for the eigenvalue $\sqrt{5}S$ of the matrix S_D, so T_1 is indeed the Jahn-Teller ground state. A complete set of the eigenvectors of this matrix (but with a different choice of axes) is given by Dunn and Bates (1995). The Ham factor $K(\mathrm{H})$ can be found from these states by the same process

---

[3]We list these as decimal numbers, which are derived from normalising the form analogous to (3.36). A list in the form of arithmetical formulas like (3.37) is much more complicated.

as before, and, as before, comes out as 2/5. Again we can note that $K(H)$ is smaller at intermediate warping strength, but probably not very small. We conclude this section by reiterating that it is necessary to take the geometric phase properly into account to correctly identify and order the energy levels, and to identify the Ham factors correctly. For instance an opposite choice of phase in the case of the icosahedral minima would replace the matrix S_I by one in which all the negative signs are removed. The eigenvectors of this matrix belong to A and H representations, with the A state lowest in energy.

3.4 CALCULATIONS FOR INTERMEDIATE COUPLING
STRENGTH

The discussion so far has been based on the adiabatic approximation in a way that assumes that the Jahn-Teller coupling is strong; in the units used here, that means that $k^2 \gg 1$. The opposite assumption, $k^2 \ll 1$, can be attacked by perturbation methods, but in between, there is the region $k^2 \sim 1$ to be charted. This intermediate region requires a numerical method. One technique that has been widely and successfully used starts with a basis of products of the uncoupled electronic and vibrational states, and after finding the matrix of the Hamiltonian in this basis, then finds the lowest eigenvalues and their corresponding eigenvectors. The fundamental difficulty with this technique is not only that a vibrator has an infinite number of energy levels, but also that when there are several modes of vibration, in this case five, the number of different states increases very rapidly with energy. We circumvent this problem to a large extent by classifying the states by their symmetry. Since the Hamiltonian will only connect states belonging to the same representation of its symmetry group, this can significantly reduce the size of the matrix. We deal with the infinite nature of the matrix by imposing an arbitrary cutoff; it turns out that for any given cutoff, the low-lying energy levels can be found with sufficient accuracy up to some value of k. It is only necessary to monitor the accuracy carefully to find that limiting k.

For T ⊗ h the existence of the extra SO(3) symmetry is a powerful tool: from a T$_1$ electronic state, all the states can be listed by angular momentum quantum numbers, and angular momentum coupling theory can be used throughout. Even more important is the way the states of the five vibrations can be classified by representations of SO(5), the group of rotations in five-dimensional space. In order to set up the matrix, we need to have enough labels for each basis state to be uniquely described, and we only need to add the phonon number n to the labels derived from the group chain SO(5)⊃SO(3) to have a complete set for the $L = 1$ matrix, which gives the ground state at all coupling strengths. For the higher angular momentum states, a few extra labels have to be added, but the method works well for the first few excited states, and it can also be extended very naturally to include electron spin. The simplicity of the labelling and subsequent setting up of the matrices comes about in the following way. The set of eigenstates of a five-dimensional harmonic oscillator with an excitation number n can be decomposed into representations of SO(5) as follows:

$$[n] \to (n, 0) + (n - 2, 0) + (n - 4, 0) + \ldots + (0, 0) \tag{3.42}$$
$$\text{or} \ + \ (1, 0),$$

where the alternative forms are for n even and n odd. The further branching rules for the reduction $SO(5) \to SO(3)$ are not quite so simple, as shown below.

$$(v, 0) \ \to \ (L) \tag{3.43}$$

$$(0, 0) \ \to \ (0)$$
$$(1, 0) \ \to \ (2)$$
$$(2, 0) \ \to \ (2) + (4)$$
$$(3, 0) \ \to \ (0) + (3) + (4) + \cdots$$
$$(4, 0) \ \to \ (2) + (4) + (5) + \cdots$$
$$(5, 0) \ \to \ (2) + (4) + (5) + \cdots$$
$$\cdots$$

where (L) indicates an angular momentum state, and the pattern for these lower angular momentum states repeats for each increase of v by 3. Clearly for these low angular momentum states, the labelling $|n, v, L\rangle$ is unique, and it is noteworthy that $L = 1$ never appears. Furthermore the T_1 electronic state is also an $L = 1$ state, so that if we want to study vibronic states with $L = 1$, we need only include the vibrational states with $L = 0, 2$. This produces a great pruning of the basis states for the matrix, and in practise makes the numerical diagonalization possible. The matrix elements are of two kinds. The uncoupled harmonic oscillator Hamiltonian produces diagonal matrix elements that are just equal to n (or $n + 5/2$ if the zero-point energy is to be included explicitly). The Jahn-Teller interaction is linear in the harmonic oscillator coordinates $\{Q_i\}$ and so only connects $[n]$ to $[n \pm 1]$. The $\{Q_i\}$ span a $(1, 0)$ irrep of SO(5), and the way these irreps combine means that $(1, 0)$ only connects $(v, 0)$ to $(v \pm 1, 0)$. The result of these selection rules is that only a few matrix elements are nonzero in any row, such that the matrix is very sparse. The actual formulas for the matrix elements are quite complicated to find, but they have been tabulated and can be found in Judd (1974) and O'Brien (1971, 1976). The size and sparseness of the matrix, together with the fact that we are most interested in the properties of the few lowest roots, is a strong indication that the best numerical method to use is the Lanczos process. (See Pooler [1984] for a brief description of this method). This is a method that is not generally approved for numerical work

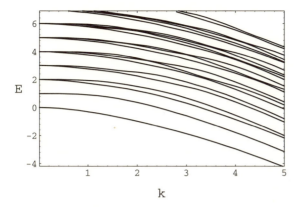

Figure 3.3. Low-lying vibronic energy levels of symmetry $L = 1$ plotted against the coupling strength k

with matrices because its bases lose their orthogonality, producing multiple copies of roots. However, it does work very fast with sparse matrices. It produces the low eigenvalues first, and it has been shown that extra copies of an eigenvalue do not appear until the first copy has converged. The infinite matrices have to be cut off at some finite value of n, which needs to be as large as possible to get good results for k as large as possible. It is the use of the Lanczos process that enables large n's to be used. Testing whether n is large enough for some given k means simply seeing what effect various cutoffs have on the results. These numerical methods have been used to follow the energies of the low-lying levels and the Ham factors $K(T_1)$ and $K(H)$ as k goes from $\ll 1$ to $\gg 1$. Here we show some results. First in Figure 3.3 we show some of the low-lying energy levels of T_1 or $L = 1$ symmetry plotted against k. Notice that the density of these levels increases with energy, and also that the spacing of these vibronic states is far from the uniform spacing to be expected from uncoupled vibrational states. Figure 3.4 shows the two lowest $L = 1$ levels and the lowest $L = 3$ level plotted against k. It is clear that the $L = 1$ level remains the lowest one for all values of k with the $L = 3$ level approaching it at strong coupling, as shown earlier. At weak coupling the ground state approaches the uncoupled electronic T_1 state, so the lowest $L = 3$ level must have a higher energy. The Ham factors are shown in Figure 3.5, which predictably shows $K(T_1)$ going from 1 to 0 over the range of k, while $K(H)$ goes from 1 to 2/5. If the electronic state is T_2 rather than T_1 the transformation (3.15) ensures that there is an identical set of matrices and resulting energy levels. The difference is that the labelling now has to interchange T_1 and T_2 wherever they occur. One important result is in the interpretation of the Ham factor, which is $K(T_1)$ in $T_1 \otimes h$ and $K(T_2)$ in $T_2 \otimes h$. Because the angular momentum transforms as T_1, we are plotting the quenching of angular momentum and the spin-orbit coupling in plotting $K(T_1)$ as it goes from 1 to 0 with increasing coupling strength. In a

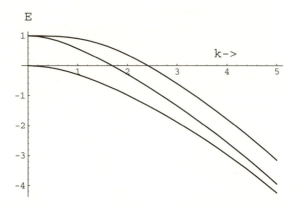

Figure 3.4. The two lowest $L = 1$ energy levels and the lowest $L = 3$ level plotted against k, showing the $L = 3$ level coming down towards the $L = 1$ ground state as k increases.

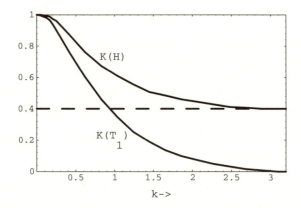

Figure 3.5. The Ham factors $K(T_1)$ and $K(H)$ in $T_1 \otimes h$, plotted against k.

T_2 state, the equivalent Ham factor is $K(T_2)$, which has nothing to do with the angular momentum. Within a T_2 state, there are no angular momentum matrix elements whatever the coupling strength, and no spin-orbit coupling.

3.5 MULTIPLE OCCUPATION OF T_1 ORBITALS

The problem of the effect of Jahn-Teller interactions in states resulting from multiple occupancy of triply degenerate t states in cubic symmetry, $p^n \otimes (\tau_2 \oplus \epsilon)$, has been extensively studied over the last twenty years in connection with the various charge states of the vacancy in silicon (V centers) by G. D. Watkins and his collaborators. This work is described in a review article (Watkins, 1986), and the connection with the similar problem in icosahedral symmetry has been pointed out by Lannoo et al. (1991). The similarity is the identity of the T ⊗ h Hamiltonian with T ⊗ $(\tau_2 \oplus \epsilon)$ in the special case of equal coupling that has already been exploited in this chapter. Auerbach et al. (1994), motivated by the need to find pairing energies to use in the discussion of superconductivity in compounds of C_{60}, have produced an extensive analysis of this problem. This section has been entitled p^n rather than t_{1u}^n because of the SO(3) invariance of the linear T ⊗ h Hamiltonian and because of the fact that all angular momentum states up to $L = 2$ remain unsplit in the icosahedral group—two reasons it is particularly convenient to use angular momentum labelling. We shall thus describe the electronic configurations as p^n and the coupled states using LS coupled terms in the usual spectroscopic notation. Setting up the Jahn-Teller interaction in bases constructed from terms rather than configurations makes it easy to allow for the inter-electron coulomb energy. In the V centers, it is the balance between the coulomb energy and the Jahn-Teller interaction that is important in determining the order of the energy levels, and we shall show how this works in icosahedral symmetry. The first thrust of the discussion that follows is in terms of the states with strong Jahn-Teller interactions, starting with the analogous discussion to that given in Section 3.3.1. Weak interaction, dealt with by numerical methods, similar to those described earlier in Section 3.4, follows later, together with the introduction of the coulomb energies that give the terms different energies.

3.5.1 The Configurations p^2 and p^4

The terms from these two configurations are ^3P, ^1D, and ^1S. The high spin state is simple, being yet another P state coupled to the h vibrations. The only extra thing we need to know is how the coupling constant in this state relates to the coupling constant k for a single electron, that is, we need to work out an extra reduced matrix element. The coupling constant turns out to be $-k$, so that the energies are as given by (3.24), but with a distortion of the opposite sign. For the low spin states we have a sixfold basis, five D states and one S state. The S state does not have any diagonal Jahn-Teller coupling, but it is coupled to the D states. The Jahn-Teller interaction matrix is

$$M_{SD} = -k \times$$

$$
\begin{bmatrix}
0 & -\sqrt{2}Q_1 & -\sqrt{2}Q_2 & -\sqrt{2}Q_3 & -\sqrt{2}Q_4 & -\sqrt{2}Q_5 \\
-\sqrt{2}Q_1 & Q_1 & \frac{1}{2}Q_2 & -Q_3 & -Q_4 & \frac{1}{2}Q_5 \\
-\sqrt{2}Q_2 & \frac{1}{2}Q_2 & \frac{1}{2}Q_1 + \frac{\sqrt{3}}{2}Q_4 & \frac{\sqrt{3}}{2}Q_5 & \frac{\sqrt{3}}{2}Q_2 & \frac{\sqrt{3}}{2}Q_3 \\
-\sqrt{2}Q_3 & -Q_3 & \frac{\sqrt{3}}{2}Q_5 & -Q_1 & 0 & \frac{\sqrt{3}}{2}Q_2 \\
-\sqrt{2}Q_4 & -Q_4 & \frac{\sqrt{3}}{2}Q_2 & 0 & -Q_1 & -\frac{\sqrt{3}}{2}Q_5 \\
-\sqrt{2}Q_5 & \frac{1}{2}Q_5 & \frac{\sqrt{3}}{2}Q_3 & \frac{\sqrt{3}}{2}Q_2 & -\frac{\sqrt{3}}{2}Q_5 & \frac{1}{2}Q_1 - \frac{\sqrt{3}}{2}Q_4
\end{bmatrix}
$$

$$(3.44)$$

where the coupling constants have been calculated as reduced matrix elements in terms of the original single-electron coupling constant k. Here and subsequently the dividing lines inside the matrices are put in only to guide the eye; here they separate the S and D bases. If this matrix is put in terms of the angular parametrization (3.5), it can be transformed to a nearly diagonal form by a sequence of orthogonal transformations. This transformation has a block diagonal form, with a 5×5 block to rotate the D bases, and a 1×1 unit matrix block for the S basis. The transformation matrix is

$$T_{SD} = A_{SD}(\alpha)B_{SD}(\gamma)C_{SD}(\theta)D_{SD}(\phi), \tag{3.45}$$

where the component matrices are given as

$$A_{SD}(\alpha) = \begin{bmatrix} 1 & 0 \\ \hline 0 & A_D(\alpha) \end{bmatrix}, \quad B_{SD}(\gamma) = \begin{bmatrix} 1 & 0 \\ \hline 0 & B_D(\gamma) \end{bmatrix},$$

$$C_{SD}(\theta) = \begin{bmatrix} 1 & 0 \\ \hline 0 & C_D(\theta) \end{bmatrix}, \quad D_{SD}(\phi) = \begin{bmatrix} 1 & 0 \\ \hline 0 & D_D(\phi) \end{bmatrix}, \tag{3.46}$$

in terms of the rotation matrices given in Appendix F. The interaction matrix $[M_{SD}(\alpha)]$ that results from this transformation is given by

$$[M_{SD}(\alpha)] = T_{SD} \times [M_{SD}] \times T_{SD}^{-1}, \tag{3.47}$$

and it takes the form

$$M_{SD}(\alpha) = -kQ \times$$

$$\begin{bmatrix} 0 & \sqrt{2}\cos\frac{3\alpha}{2} & 0 & 0 & \sqrt{2}\sin\frac{3\alpha}{2} & 0 \\ \hline \sqrt{2}\cos\frac{3\alpha}{2} & 1 & 0 & 0 & 0 & 0 \\ 0 & 0 & \cos(\alpha - \frac{\pi}{3}) & 0 & 0 & 0 \\ 0 & 0 & 0 & -\cos\alpha & 0 & 0 \\ \sqrt{2}\sin\frac{3\alpha}{2} & 0 & 0 & 0 & -1 & 0 \\ 0 & 0 & 0 & 0 & 0 & \cos(\alpha + \frac{\pi}{3}) \end{bmatrix} \tag{3.48}$$

This matrix is left incompletely diagonalized to make it obvious where extra terms would have to be put in if the ^1D and ^1S terms were not assumed degenerate in the first place. The lowest value of the lowest eigenvalue of this matrix occurs at $\alpha = 0$, and it is $-2kQ$. The eigenvector will be given by the first column of the matrix T_{SD}^{-1}, and at $\alpha = 0$ this is

$$u = \begin{bmatrix} \frac{1}{\sqrt{3}} \\ \hline \sqrt{\frac{2}{3}}v_D \end{bmatrix}, \tag{3.49}$$

where

$$
v_D = \begin{bmatrix}
\frac{1}{2}(3\cos^2\theta - 1) \\
\frac{\sqrt{3}}{2}\sin 2\theta \cos\phi \\
\sqrt{3}\sin^2\theta \cos\phi \sin\phi \\
\frac{\sqrt{3}}{2}\sin^2\theta \cos 2\phi \\
\frac{\sqrt{3}}{2}\sin 2\theta \sin\phi.
\end{bmatrix}.
\tag{3.50}
$$

This u is substituted into the product wave function Ψ (3.16), which leads to a Schrödinger equation with kinetic terms identical to those of (3.18). Of these, $\langle u|\nabla u\rangle = 0$, since u is real, and it remains to calculate the $\langle u|\nabla^2 u\rangle$ term. Operating on u (3.49) with the kinetic energy operator (3.22) gives

$$
-\frac{1}{2}\nabla^2 u = \frac{1}{6Q^2}\begin{bmatrix} 0 \\ \hline 6\sqrt{\frac{2}{3}}v_D \end{bmatrix},
\tag{3.51}
$$

so that $-\frac{1}{2}\langle u|\nabla^2 u\rangle = \frac{2}{3Q^2}$. Finally we apply the substitution (3.21) and put $\alpha = 0$ to get the vibronic Schrödinger equation below in this case.

$$
-\frac{1}{2}\left[\frac{1}{Q^2}\frac{\partial}{\partial Q}\left(Q^2\frac{\partial}{\partial Q}\right) + \frac{1}{Q^2\sin\alpha}\frac{\partial}{\partial\alpha}\left(\sin\alpha\frac{\partial}{\partial\alpha}\right)\right.
\tag{3.52}
$$
$$
\left. + \frac{1}{Q^2\sin^2\alpha}\frac{\partial^2}{\partial\gamma^2} - Q^2 + 4kQ\right]\psi
$$
$$
+\frac{1}{Q^2}\left[\frac{2}{3} - \frac{1}{3}\right]\psi - \frac{1}{6Q^2}\left[\frac{1}{\sin\theta}\frac{\partial}{\partial\theta}\left(\sin\theta\frac{\partial}{\partial\theta}\right) + \frac{1}{\sin^2\theta}\frac{\partial^2}{\partial\phi^2}\right]\psi
$$
$$
= E\psi.
$$

Here $f(\alpha)$ in (3.22) takes its $\alpha = 0$ value, 2/3. In the same way as (3.23), this equation represents a three-dimensional harmonic oscillator with a rotator, and the energies of the low-lying eigenvalues are

$$
E_{2,4} = -2k^2 + \frac{3}{2} + \frac{1}{12k^2} + \left(\frac{1}{24k^2}\right)L(L+1),
\tag{3.53}
$$

where L is a positive integer or 0. The phase changes over the $\alpha = 0$ surface in this case limit L to be an even integer or 0, in contrast to the situation in the case of p^1. This is because the electronic basis (3.49) is even under inversion, and the whole vibronic wave function must also be even under inversion. The two lowest states are thus S and D states as are the uncoupled electronic states (see Figure 3.6 at Section 3.5.4).

3.5.2 The Configuration p^3

The terms from this configuration would be ^4S, ^2D, and ^2P before coupling. Here the high spin state is trivial, having no Jahn-Teller coupling. For the low spin states we start by finding reduced matrix elements of the Jahn-Teller coupling within and between ^2D and ^2P. The result is that the only nonzero Jahn-Teller coupling is between ^2D and ^2P, and the matrix elements coupling P to P or D to D are 0. The interaction matrix in the basis of P and D states is found using vector coupling coefficients to be

$$M_{PD} = \frac{\sqrt{3}}{2} k \left[\begin{array}{c|c} 0 & \mathcal{M}_{PD} \\ \hline \mathcal{M}^T_{PD} & 0 \end{array} \right], \tag{3.54}$$

where

$$\mathcal{M}_{PD} = \left[\begin{array}{cccccc} \sqrt{3}Q_5 & Q_3 & -Q_2 & Q_5 & -Q_4 - \sqrt{3}Q_1 \\ -\sqrt{3}Q_2 & -Q_4 + \sqrt{3}Q_1 & Q_5 & Q_2 & -Q_3 \\ 0 & -Q_5 & 2Q_4 & -2Q_3 & Q_2 \end{array} \right] \tag{3.55}$$

and $\mathcal{M}^T_{PD}$ is the transpose of (3.55). The lack of diagonal blocks in this coupling matrix was noted in another context by Anderson et al. in 1984, and was discussed in terms of quasi-spin quantization by Ceulemans (1994). This matrix is transformed by a series of rotations in the angles (γ, θ, ϕ) and the 8×8 rotation matrix is made up in block-diagonal form from matrices given in Appendix F as follows.

$$T_{PD} = \left[\begin{array}{c|c} B_P(\gamma) & 0 \\ \hline 0 & B_D(\gamma) \end{array} \right] \left[\begin{array}{c|c} C_P(\theta) & 0 \\ \hline 0 & C_D(\theta) \end{array} \right] \left[\begin{array}{c|c} D_P(\phi) & 0 \\ \hline 0 & D_D(\phi) \end{array} \right]. \tag{3.56}$$

The result is

$$M_{PD}(\alpha) = T_{PD} \times M_{PD} \times T^{-1}_{PD} = \sqrt{3}kQ \left[\begin{array}{c|c} 0 & \mathcal{M}_{PD}(\alpha) \\ \hline \mathcal{M}^T_{PD}(\alpha) & 0 \end{array} \right], \tag{3.57}$$

where

$$\mathcal{M}_{PD}(\alpha) = \left[\begin{array}{ccccc} 0 & 0 & 0 & 0 & -\sin(\alpha + \frac{\pi}{3}) \\ 0 & -\sin(\alpha - \frac{\pi}{3}) & 0 & 0 & 0 \\ 0 & 0 & \sin\alpha & 0 & 0 \end{array} \right], \tag{3.58}$$

and $\mathcal{M}^T_{PD}(\alpha)$ is its transpose. At this stage the matrix is still in block form, so that different energies for the ^2D and ^2P states could be inserted on the diagonal. With these states degenerate, the eigenvalues are

$$E = \sqrt{3}kQ \left\{ \pm \sin\left(\alpha + \frac{\pi}{3}\right), \pm \sin\left(\alpha - \frac{\pi}{3}\right), \pm \sin\alpha, 0, 0 \right\}. \tag{3.59}$$

All of these roots that are not identically 0 have the same minimum energy for some choice of α, and we choose to take $\sin \alpha = 1$. The fact that this lies outside the original region $0 < \alpha < \frac{\pi}{3}$ does not matter; we are just looking at a different copy of the phase space. With this choice of α, the $\{Q_i\}$ are given by

$$Q_1 = Q\left(\frac{1}{2}\sqrt{3}\sin^2\theta\cos 2\gamma\right), \tag{3.60}$$

$$Q_2 = Q\left(-\frac{1}{2}\sin 2\theta\cos\phi\cos 2\gamma + \sin\theta\sin\phi\sin 2\gamma\right),$$

$$Q_3 = Q\left(\frac{1}{2}(1+\cos^2\theta)\sin 2\phi\cos 2\gamma + \cos\theta\cos 2\phi\sin 2\gamma\right),$$

$$Q_4 = Q\left(\frac{1}{2}(1+\cos^2\theta)\cos 2\phi\cos 2\gamma - \cos\theta\sin 2\phi\sin 2\gamma\right),$$

$$Q_5 = Q\left(-\frac{1}{2}\sin 2\theta\sin\phi\cos 2\gamma - \sin\theta\cos\phi\sin 2\gamma\right),$$

while the eigenstate for the lowest energy comes out as

$$u = \frac{1}{\sqrt{2}}\begin{bmatrix} \sin\theta\cos\phi \\ \sin\theta\sin\phi \\ \cos\theta \\ \hline \frac{\sqrt{3}}{2}\sin^2\theta\sin 2\gamma \\ -\cos\theta\sin\theta\cos\phi\sin 2\gamma - \sin\theta\sin\phi\cos 2\gamma \\ \frac{1}{2}(1+\cos^2\theta)\sin 2\phi\sin 2\gamma - \cos\theta\cos 2\phi\cos 2\gamma \\ \frac{1}{2}(1+\cos^2\theta)\cos 2\phi\sin 2\gamma + \cos\theta\sin 2\phi\cos 2\gamma \\ -\cos\theta\sin\theta\sin\phi\sin 2\gamma + \sin\theta\cos\phi\cos 2\gamma \end{bmatrix}. \tag{3.61}$$

To find the strong coupling pseudo-rotational states in this case, we must use the appropriate form for the vibrational kinetic energy, which is (3.19) at $\alpha = \pi/2$. In order to standardize the harmonic oscillator part of the operator, this time we take a factor $1/\sqrt{Q^3 \sin 3\alpha}$ out of ψ before setting $\alpha = \pi/2$ and get

$$\mathcal{H}_{KE} = -\frac{1}{2}\left[\frac{1}{Q}\frac{\partial}{\partial Q}\left(Q\frac{\partial}{\partial Q}\right) + \frac{1}{Q^2}\frac{\partial^2}{\partial\alpha^2} + \frac{9}{4Q^2}\right]$$
$$-\frac{1}{8Q^2}\left[4\lambda_x^2 + 4\lambda_y^2 + \lambda_z^2\right]. \tag{3.62}$$

The operator on the second line is a version of the Hamiltonian of the symmetric top. The energies and eigenfunctions of the symmetric top and their relationship to the representations of finite rotations are set out in Chapter 4 of Edmonds (1960). The eigenfunctions are

$$\mathcal{D}^{(L)}_{M,K}(\phi,\theta,\gamma) = e^{i(M\phi+K\gamma)}d^{(L)}_{M,K}(\theta), \tag{3.63}$$

where L, M, and K are integers, $L \geq |M|$, and $L \geq |K|$. With the effective moments of inertia in (3.62), the kinetic energy eigenvalues are

$$\frac{1}{2Q^2}\left[L(L+1) - \frac{3}{4}K^2 \right].\tag{3.64}$$

To find the $\nabla^2 u$ term, we notice that the components of the eigenvector (3.61) is made up of these symmetric top eigenstates. If it is written

$$u = \frac{1}{\sqrt{2}}\begin{bmatrix} v_1 \\ v_2 \end{bmatrix},\tag{3.65}$$

then v_1 is composed of states with $L = 1$, $K = 0$ while v_2 has $L = 2$, $K = 2$, so using (3.64) gives

$$-\frac{1}{2}\nabla^2 u = \frac{1}{\sqrt{2}}\frac{1}{2Q^2}\begin{bmatrix} 2v_1 \\ 3v_2 \end{bmatrix}.\tag{3.66}$$

This leads to

$$-\left\langle u \left| \frac{1}{2}\nabla^2 u \right. \right\rangle = \frac{1}{2Q^2}\frac{5}{2},\tag{3.67}$$

a term in the Schrödinger equation (see the development after equation [3.50]). Thus the vibronic Schrödinger equation is

$$-\frac{1}{2}\left[\frac{1}{Q}\frac{\partial}{\partial Q}\left(Q\frac{\partial}{\partial Q} \right) + \frac{1}{Q^2}\frac{\partial^2}{\partial \alpha^2} - Q^2 + 2\sqrt{3}kQ \right]\psi \tag{3.68}$$
$$+\frac{1}{Q^2}\left[\frac{5}{4} - \frac{9}{8} \right]\psi - \frac{1}{8Q^2}\left[4\lambda_x^2 + 4\lambda_y^2 + \lambda_z^2 \right]\psi = E\psi.$$

The top row of this equation represents a two-dimensional harmonic oscillator with a displaced origin, and the second row a symmetric top, also with the energy origin displaced. Using the value of Q at the minimum, $Q = \sqrt{3}k$, the energies come out as

$$E_3 = -\frac{3}{2}k^2 + 1 + \frac{1}{24k^2} + \frac{1}{6k^2}\left[L(L+1) - \frac{3}{4}K^2 \right].\tag{3.69}$$

The electronic eigenstate (3.61) is seen to be invariant under (3.25) and to change sign under (3.26), and the pseudo-rotational eigenstates must change sign with the electronic eigenstate to preserve invariance. Thus invariance under (3.25) requires K to be even, and the sign change under (3.26) requires the pseudo-rotational eigenstates to be

$$\Phi_{rot} = d^{(L)}_{M,K}(\theta)e^{iM\phi} \times \begin{cases} \cos K\gamma & \text{for L odd;} \\ \sin K\gamma & \text{for L even.} \end{cases}\tag{3.70}$$

The lowest state has ($L = 1$, $K = 0$), followed closely in energy by ($L = 2$, $K = 2$); next comes ($L = 4$, $K = 4$), followed by ($L = 3$, $K = 2$). There are no $L = 0$ states, and $K = 0$ only occurs with odd L. Unlike the other configurations, where motion on the $\alpha = 0$ surface corresponds to a rotating spheroidal ($3z^2 - r^2$) distortion, the rotating distortion on the $\alpha = \pi/2$ surface is of ($x^2 - y^2$) type, which is not axially symmetric and so requires the three angles (γ, θ, ϕ) to describe its orientation. This is why in the case of p^3 alone, the ground vibronic states are functions of three angles, not two. This distinction between the distortions for $p^2 \otimes h$ and $p^3 \otimes h$ is borne out in the simulations of Fagerström and Stafström (1993).

3.5.3 Numerical Work on p^n

All of these systems can be set up as matrices in the uncoupled states for numerical diagonalization, using the SO(5) group for handling the phonon excitations as described in Section 3.4, though here the coupling scheme and hence the matrix structure is more complicated. For the singlet states in p^2 and p^4, two different matrices must be set up, one in bases of final S symmetry and one of final D symmetry. Both will contain bases originating from the electronic ^1S and ^1D states. For the S matrix we need vibrational states corresponding to $L = 0$ to go with ^1S, and $L = 2$ to go with ^1D. For the D matrix, vector coupling rules give $L = 2$ with ^1S and $L = 0, 2, 3, 4$ with ^1D. (As remarked before, $L = 1$ never appears in the vibrational state). Although the coupling scheme is complicated, the essential condition that the labelling is unique holds good. Every basis can be labelled by n, the phonon excitation number, with v labelling the SO(5) representation, L labelling the SO(3) representation, and an index specifying the electronic state. The selection rules for the matrix elements, which connect states with n and v only differing by one unit, keeps the matrix sparse. The matrix elements are products of vector coupling coefficients and fractional parentage coefficients, as detailed by O'Brien (1976) and the fractional parentage coefficients listed in that paper go to high enough L for the present purpose. In order to track the lowest energy level in each case, the Lanczos process is very suitable (see Section 3.4). Using this method it is possible to employ a large enough matrix to verify the asymptotic forms of the two lowest energies calculated for strong coupling (3.53). The matrix for p^3 is much simplified by the lack of Jahn-Teller coupling within ^2P and within ^2D. States with even numbers of phonons associated with ^2P will only couple to states with odd numbers of phonons associated with ^2D, and vice versa. In particular as no S states can be produced from electronic P states with vibrations of this symmetry, no S states will change energy with Jahn-Teller coupling strength. For the P matrix, vector coupling rules give ^2P with $L = 0, 2$ and ^2D with $L = 2, 3$. For the D matrix they give ^2P with $L = 2, 3$ and ^2D with $L = 0, 2, 4$ (the expected $L = 3$ term goes out because of a chance 0 in a $6j$ symbol). Again the existing techniques and fractional parentage coefficients will set up the matrix, and Lanczos will find the lowest eigenvalues. The strong coupling asymptotic lowest energies (3.69) can be verified by this calculation.

3.5.4 Term Splittings and Energy Ordering

The Coulomb energies for the terms of all these configurations are shown in the following tables:

$p^2(p^4)$	Energy
^1S	$(6)F_0 + 10F_2$
^1D	$(6)F_0 + F_2$
^3P	$(6)F_0 - 5F_2$

and

p^3	Energy
^2P	$3F_0$
^2D	$3F_0 - 6F_2$
^4S	$3F_0 - 15F_2,$

$$(3.71)$$

where F_0 and F_2 are certain integrals of the Coulomb interaction within charge distributions depending on the p wave functions. Tables of term energies in this form are familiar as the actual energies of the atomic configuration, but they can be derived more generally using symmetry principles. An atomic p state is a special case of a general t_{1u} molecular orbital. A treatment of this and similar problems can be found in Sugano et al. (1970). The quantities F_0 and F_2 are integrals of the Coulomb interaction within certain charge distributions depending on the molecular orbital wave functions, and as self-energies they are intrinsically positive. The important result here is that all the term splittings within each configuration depend on a single parameter, F_2, that is known to be positive. The effect of including these term splittings at strong Jahn-Teller coupling can be seen by including them in the various interaction matrices. In p^2 and p^4 the Jahn-Teller depression of the ^3P state is $1/4$ of that in the combined singlet states, so that at strong Jahn-Teller coupling, the lowest ^4P state is above the singlet states; we have seen that of the singlet states, ^1S is below ^1D. There is thus a complete reversal of the ordering of the states as the Jahn-Teller coupling strength is turned up. The effect can be tracked quantitatively by including the term energies on the diagonals of the matrices described in Section 3.5.3. An example of such a calculation with and without the term energies is shown in Figures 3.6 and 3.7. Figure 3.7 illustrates the reversal of ordering of the states. For p^3 there is an uncoupled ^4S state with no Jahn-Teller interaction. Because of the term splitting between the ^2D and ^2P states themselves, the Jahn-Teller effect will only come in at second order, so it is effectively quenched if it is weaker than the term splitting. A strong Jahn-Teller interaction puts the ^2P state marginally below the ^2D state, and both well below the ^4S, so again a strong Jahn-Teller interaction reverses the order of the states. Examples of calculations for p^3 are shown in Figures 3.8 and 3.9.

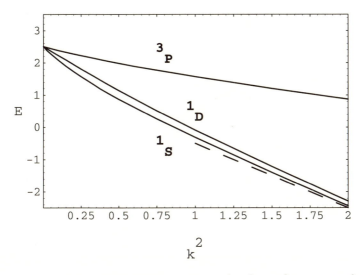

Figure 3.6. Calculated energy levels for the lowest ^1D, ^1S and ^3P states of p^2 or p^4 without any term splitting energy, plotted against k^2. Energy is in units of $\hbar\omega$. The dashed line shows the asymptotic energy.

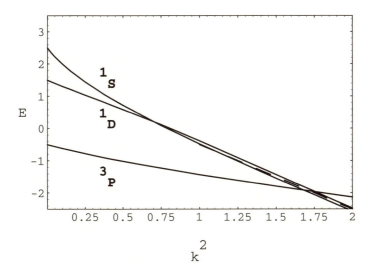

Figure 3.7. As Figure 3.6, but with a term splitting corresponding to $F_2 = \hbar\omega/3$ included. At $k^2 = 0$ the lowest ^1S state is a one-phonon excitation from ^1D; this is **not** the ^1S electronic state which, at $k^2 = 0$, is at $E = 4.5\hbar\omega$.

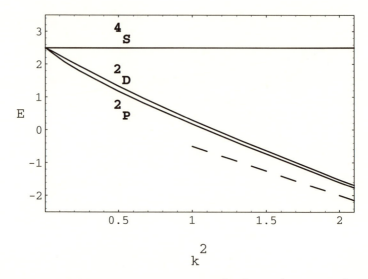

Figure 3.8. Calculated energy levels for the lowest ^2P, ^2D and ^4S states of p^3 without any term splitting energy, plotted against k^2. Energy is in units of $\hbar\omega$. The dashed line shows the asymptotic energy.

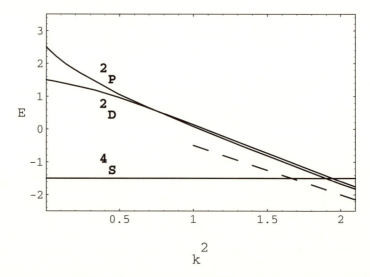

Figure 3.9. As Figure 3.8, but with a term splitting corresponding to $F_2 = \hbar\omega/3$ included. At $k^2 = 0$ the lowest ^2P state is a one-phonon excitation from ^2D; this is **not** the ^2P electronic state which at $k^2 = 0$ is at $E = 3.5\hbar\omega$.

3.6 OPTICAL ABSORPTION SPECTRA

In the absence of a Jahn-Teller interaction, the "normal" characteristic of vibrational structure on an electronic transition in a molecule is that the energies can be analyzed in terms of the regular series of excitations $n\hbar\omega$ of the oscillators concerned. In a solid the multiplicity of modes produces a "normal" spectral profile that is a band of Gaussian shape, with or without a detectable zero-phonon line. The signature of a Jahn-Teller interaction in a line spectrum is that the sequence of energies is no longer uniformly spaced, and the intensities vary with position in an irregular way. The signature in a solid is that the bands depart markedly from the "normal" Gaussian shape and may develop quite sharp features.

3.6.1 Molecular Spectra

We concentrate on allowed transitions, such as from s to p electronic states, and, as we shall see, the effects are seen in absorption when the excited p state suffers a Jahn-Teller interaction. We assume that the whole of the transition probability arises from the electronic properties, and that the vibrations are only effective in altering the transition matrix element by an overlap. To be precise, if the ground state wave function is written

$$\Psi_0 = \psi_0(\mathbf{Q})|s\rangle, \tag{3.72}$$

and an excited state wave function is written

$$\Psi_1 = \psi_x(\mathbf{Q})|p_x\rangle + \psi_y(\mathbf{Q})|p_y\rangle + \psi_z(\mathbf{Q})|p_z\rangle, \tag{3.73}$$

and if for instance the nonzero transition probability for a particular polarization of dipole radiation is between $|s\rangle$ and $|p_x\rangle$ and is $\mathcal{P}$, then the transition probability between the above two states is

$$|\langle\psi_0(\mathbf{Q})|\psi_x(\mathbf{Q})\rangle|^2\mathcal{P}. \tag{3.74}$$

In the system $s \rightarrow p$, or $A \rightarrow T_1$, where the angular momentum is a good quantum number, it follows that the allowed transitions are only to the excited $L = 1$ states, so the energies of the allowed transitions are given by a section through Figure 3.3 at a constant k. It is clear that the spacing is far from uniform, as is illustrated by Figures 3.10 and 3.11, which show the calculated spectra for two different values of k. In these figures the relative intensities of the lines are shown, having been found by calculating the squares of the overlaps. Each line in the spectra has been convoluted with a narrow Gaussian to make the plotting easier. There are many other lines too weak to appear in the figures. One feature that is obvious in these two spectra is that the low energy end

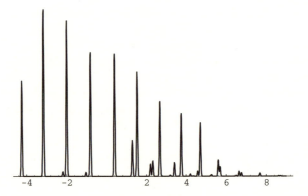

Figure 3.10. The calculated line spectrum for a transition to a set of $T \otimes h$ vibronic states with coupling given by $k = 5$, produced as described in the text.

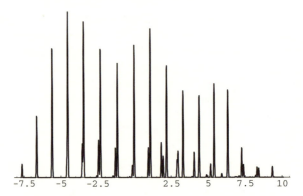

Figure 3.11. The calculated line spectrum for $k = 5\sqrt{2}$

begins to resemble the "normal" spectrum that would arise without any Jahn-Teller coupling, while the high energy end is much more complicated. This is because the lower energy levels are associated with the single lowest APES with its approximately paraboidal shape at the minimum, while at higher energies, transitions to all the APESs are superimposed. A major drawback to the use of the calculations reported here is that they have not been extended to cover the possibility of vibrational modes of several different frequencies engaging simultaneously in the Jahn-Teller interaction. All we can say is that the spectrum would be immensely more complicated than the ones shown here at weak and intermediate coupling strengths, and that the way to approach such a problem would be through perturbation theory. As we shall see later, the multimode problem is, surprisingly, much more tractable in a solid, where there is a very large number of vibrational mode frequencies. The numerical diagonalization procedure for the spectra is necessarily more involved than that required for determining only the lowest energy states. A Lanczos process is used, but because of the propensity to find multiple copies of the eigenvalues, it is used in a form devised by Parlett and Reid (1980). In this program every converged eigenvalue is isolated as soon as it converges, and its spectral neighborhood is never revisited. In this way all the eigenvalues in any specified energy region can be found without duplication.

3.6.2 Absorption Bands in Solids: The Cluster Model

A solid is an extreme example of a system where there are many modes present of every symmetry type. This makes it impossible to calculate the position of individual energy levels—as would be needed if the spectrum were resolved—but the presence of so many coupled modes makes it unlikely that the spectrum is in fact resolved. For the unresolved band shape, we can use the cluster model described in Section 6.1.2. Here the multimode Hamiltonian is replaced by a single-mode or cluster Hamiltonian such that all the moments of the band are correctly given to highest order in the cluster coupling constant k_{eff}. This means that the overall unresolved band shape can be calculated to a good approximation in complete ignorance of the detailed positions of the lines. One inevitable source of broadening will be coupling with the many modes of A symmetry, the *symmetric* modes. These are the modes that produce a Gaussian band shape for bands in solids, and we represent this broadening by convoluting the calculated Jahn-Teller band shape with a Gaussian-shape function. In real life this broadening may be large enough to wash out the Jahn-Teller structure altogether, but there is little point in displaying such smoothed band shapes; instead we show bands that have been convoluted with a Gaussian that is only just large enough to remove the line structure, and leave it to the imagination of the reader to visualise the effects of more smoothing. In Figure 3.12 such a band shape is shown. This corresponds to stronger coupling than Figures 3.10 and 3.11 and the Gaussian is broader, but otherwise it is produced in the same way. The only difference is that for stronger coupling, the matrix size increases drastically, and the necessary computing effort increases even faster. In this figure the coupling strength is large enough to show clearly the three peaks that are characteristic of this system. These peaks can be accounted for by applying the strong coupling Franck-Condon approximation to the transition. In this approximation we confine our attention to "vertical" transitions from an uncoupled ground state, assumed to be the ground state of the five-dimensional harmonic oscillator, to all the APESs of the Jahn-Teller coupled excited state. Wherever a vertical or constant $\mathbf{Q}$ transition arrives on an APES, a contribution is made to the intensity of the band at the appropriate energy, and of magnitude inversely proportional to $\partial \mathbf{Q}/\partial E$. The three peaks then correspond to vertical transitions arriving on the three APESs. If, as is normal in this approximation, only the linear Jahn-Teller interaction is included,

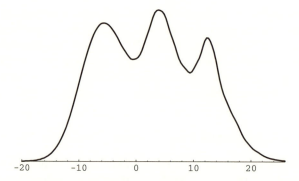

Figure 3.12. Smoothed band shape for transition from an uncoupled A state into a coupled T state, coupling strength $k = 14$, showing three peaks corresponding to the three APES's. The horizontal scale is in units of $\hbar\omega$.

the energies of the surfaces are symmetrical with respect to the uncoupled energy, and the band is symmetrical about its center. The value of k chosen for (3.12) is too small to show a completely symmetric band. To produce the symmetric band corresponding to this approximation, one can use a numerical method, introduced by Cho in 1968, which is based on adding up contributions from a random distribution of points in $\mathbf{Q}$ space. This technique is used to produce the band shapes in Section 4.6.

3.7 THE INTRODUCTION OF SPIN-ORBIT COUPLING

An electronic p or T_1 state is likely to have active spin-orbit coupling, so we should be prepared to include a term in the Hamiltonian deriving from $\lambda\mathbf{L.S}$. As would be expected, the result of including this operator is quite different according to whether the effect of spin-orbit coupling is large or small compared with the effect of the Jahn-Teller coupling.

3.7.1 $\lambda \ll E_{JT}$

With $\lambda \ll E_{JT}$ we first confine our attention to the lowest vibronic triplet. The angular momentum operator transforms as does the T_1 representation of the group, and consequently $\lambda\mathbf{L.S}$ can be replaced by $K(T_1)\lambda\mathbf{L.S}$ within the T ground state. It is clear from Figure 3.5 that $K(T_1)$ reduces the effect of the spin-orbit coupling very strongly, and the value of the effective λ to be used can be readily found for any k. If we are interested in properties of excited states, as for instance in molecular spectra, things are more complicated. It becomes necessary to set up and diagonalize the Hamiltonian in the basis of uncoupled states, using the SO(3) coupling scheme in a similar way to the

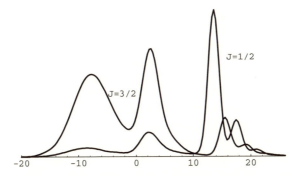

Figure 3.13. Band shape for transition into a coupled $^2T_1$ (or ^2P) state with spin-orbit coupling, $k = 14$, λ negative. The two spectra, for $J = 1/2$ and $J = 3/2$, are plotted on the same horizontal scale, which is in units of $\hbar\omega$.

method described in Section 3.4. The angular momentum notation can still be used because the spin-orbit interaction is invariant under SO(3), and so does not reduce the symmetry of the Hamiltonian. There are now two disjoint matrices, for $J = 1/2$ and $J = 3/2$, and the $J = 3/2$ matrix is larger than the $L = 1$ matrix used previously. However, the calculation can be done quite easily to predict molecular spectra if only one frequency of vibration is involved.

3.7.2 λ Comparable to or Larger than E_{JT}

If λ is comparable to or larger than E_{JT}, then it is more instructive to think first of the $J = 1$ state being coupled with the spin to give two electronic states, $J = 1/2$ and $J = 3/2$. Because all states now have half-integral spin, they all have Kramers degeneracy, so there are still only three distinct APESs. These can be best found by applying a further rotation to the matrix (3.10), the rotation being chosen so that the spin-orbit coupling is diagonal. The resulting matrix is

$$\begin{bmatrix} -\frac{1}{2}kQ\cos 3\alpha + \frac{1}{2}\lambda & \frac{1}{2}kQ\sin 3\alpha & \frac{1}{\sqrt{2}}kQ \\ \frac{1}{2}kQ\sin 3\alpha & \frac{1}{2}kQ\cos 3\alpha + \frac{1}{2}\lambda & 0 \\ \frac{1}{\sqrt{2}}kQ & 0 & -\lambda \end{bmatrix}. \tag{3.75}$$

When λ is large, the single (Kramers degenerate) APES from $J = 1/2$ is well separated from the two (Kramers degenerate) APESs from $J = 3/2$, and only the latter suffer a Jahn-Teller splitting. Figure 3.13 shows a band shape for strong coupling, with λ negative and large enough to compete with the Jahn-Teller energy. This was produced using the same technique of matrix diagonalization as was used for Figure 3.12. The bands derived from the two disjoint matrices, $J = 1/2$ and $J = 3/2$, are both shown in the figure. The effect of the three APESs is clearly visible in the three main peaks, as is the mixing of states on the different APESs by the non–block-diagonal part of (3.75). A further interesting effect is clearly shown in Figure 3.13, and this is the structure to be seen on the high energy end of the band. This structure consists of a set of resonances. Their spacing is greater than $\hbar\omega$, and they have nothing to do with the uncoupled phonon levels; they only appear and sharpen up when the coupling is strong. Their appearance is the result of a "centrifugal" term in the kinetic energy, related to the extra rotational symmetry of the Hamiltonian. The result of applying the transformation (3.11) to the kinetic energy operator produces a set of terms to be added to the matrix (3.75), all of which include a factor $1/(Q\sin(3\alpha))^2$. In the absence of this term, the $J = 1/2$ level with no Jahn-Teller coupling is associated with an APES that is just the potential of a five-dimensional oscillator. Only the transition to the oscillator ground state would be allowed in this case, and the band should reduce to a delta function. The centrifugal term adds a very anharmonic contribution, and the position of the resonances can be predicted by solving the Schrödinger equation in this strange-shaped well. The selection rules are different for $J = 1/2$ and $J = 3/2$, which accounts for the displacement of the resonances in the respective bands. The existence of this type of resonance was first pointed out by Sloncjewski (1963), and their appearance on the high energy end of a band is a very good indicator of the existence of strong Jahn-Teller coupling. The work from which this account was extracted was done for the system $T \otimes (\tau_2 \oplus \epsilon)$ and is reported in much more detail by Chancey and O'Brien (1989).

4

Electronic Quartets and $G \otimes (g \oplus h)$

4.1 INTRODUCTION

With quartet electronic states we reach the point where the icosahedral group breaks new ground in Jahn-Teller studies, because under no lower symmetry can we expect to find a fourfold degenerate state that is not also Kramers degenerate. As indicated in the chapter heading, the symmetric square of the G representation contains both G and H, so we have to consider the variation of four vibronic energy levels in a nine-dimensional coordinate space. Unlike the $T \otimes h$ systems, here we have a Hamiltonian that is not invariant under a group higher than I_h except for that choice of couplings that makes the effect of h and g vibrations "equal" to produce an overall symmetry of SO(4).

In many ways this system is similar in cubic symmetry to $T \otimes (e \oplus t_2)$, where an electronic triplet state can be coupled to two different modes of vibration. There, coupling to one or another of the modes alone produces a Hamiltonian of cubic symmetry, but one particular set of parameters, corresponding to "equal" coupling, increases the symmetry to SO(3). The approach used there is to consider as fully as possible the systems with one or another of the vibrations acting on its own, and then to use the equal coupling solutions to help in interpolating between the two. Following the same program here we start with $G \otimes g$, with g vibrations only. We proceed to $G \otimes h$ and then to the full system, which can be done in three ways: g with an h perturbation, h with a g perturbation, and the equal coupling case. As with Chapter 3, most of this chapter assumes strong Jahn-Teller coupling. This is because the problem can largely be solved at strong coupling while the intermediate coupling strengths of real systems have so far proved too hard to handle in Hamiltonians with this number of degrees of freedom, having to be estimated by interpolation between strong and weak coupling. Substantial work on this system has been done by Ceulemans and Fowler (1989), who give a very full account of the group theoretical analysis of these systems and of the characterization of the lowest APES. In this chapter we draw on this work and also on work published by Cullerne and O'Brien (1994). These authors differ in their choice of variables for describing SO(4) symmetry, and we follow the choice of Cullerne and O'Brien.

4.2 $G \otimes g$

The Hamiltonian for $G \otimes g$ can be written

$$\mathcal{H} = -\frac{1}{2}\hbar^2 \sum_{i=1}^{4} \frac{\partial^2}{\partial g_i^2} + \frac{1}{2}\omega_g^2 \sum_{i=1}^{4} g_i^2 + M^G(g), \tag{4.1}$$

where the first two terms represent the kinetic and potential energies of the four vibrations (with the mass equal to one for convenience). The Jahn-Teller interaction is

$$M^G(g) = k_g^G \omega_g \sqrt{\hbar\omega_g} \begin{bmatrix} -g_3 & -g_4 & -g_1 + g_3 & -g_2 - g_4 \\ -g_4 & g_3 & g_2 - g_4 & -g_1 - g_3 \\ -g_1 + g_3 & g_2 - g_4 & g_1 & -g_2 \\ -g_2 - g_4 & -g_1 - g_3 & -g_2 & -g_1 \end{bmatrix}. \tag{4.2}$$

This matrix is found by the methods described in Chapter 2 and is listed in Appendix E, where a set of bases that span a set of G electronic states are also given. The notation has been altered a little from the previous chapters. Because there are two sets of vibrational coordinates to be considered, belonging to the two different symmetries g and h, the coordinates here are labelled g_i and h_i instead of the slightly ambiguous Q_i used so far. Also the extra factor ω_g^2 has been included in the elastic restoring force to allow for the different frequencies of different modes, and all normal coordinates are scaled to unit mass. The extra factors $\omega_g\sqrt{\hbar\omega_g}$ in the interaction matrix, $M^G(g)$, ensure that the coupling coefficient squared, k_g^2, is the ratio of the Jahn-Teller energy to $\hbar\omega_g$ as usual. At strong coupling the four roots of the matrix (4.2) added onto the elastic energy $\frac{1}{2}\omega_g^2 \sum_{i=1}^{4} g_i^2$ give the four APESs. It is the geometry of these surfaces that determines the properties of the ground states.

4.2.1 The Method of Öpik and Pryce

The first thing to do to characterize these surfaces is to find the minima and other stationary points of the APES. To do that, we cannot do better than to follow a procedure devised by Öpik and Pryce (1957) that solves a set of simultaneous equations to find the electronic bases at stationary points. If we let **a**, a vector with components a_i, represent the eigenvector of the matrix $M^G(g)$, then the eigenvalue equation for the energy, E, of the APES is

$$M^G(g)\mathbf{a} + \frac{1}{2}\sum_i \omega_g^2 g_i^2 \mathbf{a} = E\mathbf{a}, \tag{4.3}$$

or if $M^G(g)$ is put in terms of the **U** matrices (E.10), it reads

$$\sum_i \left(k_g^G \omega_g \sqrt{\hbar\omega_g} \mathbf{U}^G(G, i) g_i + \frac{1}{2}\omega_g^2 g_i^2 \right) \mathbf{a} = E\mathbf{a}. \tag{4.4}$$

By multiplying on the left with $\mathbf{a}^T$, the transpose of $\mathbf{a}$, and making use of the normalization

$$\mathbf{a}^T \mathbf{a} = 1, \tag{4.5}$$

(4.4) becomes

$$\sum_i \left(k_g^G \omega_g \sqrt{\hbar \omega_g} \mathbf{a}^T \mathbf{U}^G(G, i) \mathbf{a} g_i + \frac{1}{2} \omega_g^2 g_i^2 \right) = E. \tag{4.6}$$

Next we apply the conditions for a stationary point,

$$\frac{\partial E}{\partial g_i} = 0, \qquad i = 1, \dots, 4, \tag{4.7}$$

to (4.4) to get

$$k_g^G \omega_g \sqrt{\hbar \omega_g} \mathbf{a}^T \mathbf{U}^G(G, i) \mathbf{a} + \omega_g^2 g_i = 0, \qquad i = 1, \dots, 4, \tag{4.8}$$

and the set of equations (4.8) together with (4.4) and (4.5) gives a set of nonlinear equations in the four a_i and E. On the face of it, there also ought to be terms in $(\partial / \partial g_i) \mathbf{a}$ included in (4.8), but (4.5) ensures that the vector $(\partial / \partial g_i) \mathbf{a}$ is orthogonal to $\mathbf{a}$, and this fact, together with (4.3) ensures that

$$\left(\frac{\partial}{\partial g_i} \mathbf{a}^T \right) M^G(g) \mathbf{a} = 0 = \mathbf{a}^T M^G(g) \left(\frac{\partial}{\partial g_i} \mathbf{a} \right). \tag{4.9}$$

These equations (4.8), (4.4), and (4.5), which have to be solved to find the a_i's and the energies at the stationary points, are a bit heavy to be solved by hand, but some solutions can be found with the help of algebraic computing packages (we have used Mathematica). Once the a_i's are found, the g_i's follow immediately from (4.8). Also, because the whole Hamiltonian has icosahedral symmetry, once any solution has been found, it can be used to find all the others related by symmetry by applying appropriate rotations. Once the stationary points are found, it is still necessary to characterize them as minima or saddle points[1], which can be done by the use of second-order perturbation around them. To show how this can be done, we shall look at a simple solution of the Öpik and Pryce equations for $G \otimes g$, which can be found by inspection. This solution is

$$\mathbf{a} = \left(\frac{-1}{\sqrt{2}}, 0, \frac{1}{\sqrt{2}}, 0 \right), \tag{4.10}$$

which, when applied in (4.8), results in

$$(g_1, g_2, g_3, g_4) = k_g^G \sqrt{\frac{\hbar}{\omega_g}} \left(-\frac{3}{2}, 0, \frac{3}{2}, 0 \right) \tag{4.11}$$

---

[1] There are no maxima in E because any turning point must be minimal with respect to a change in the magnitude of the distortion, $\sqrt{\sum_i g_i^2}$.

at an energy

$$E = -\frac{9}{4}k_g^{G2}\hbar\omega_g. \tag{4.12}$$

At this set of values of the g_i's, the other three eigenvalues of $M^G(g)$ are degenerate at an energy $4k_g^{G2}\hbar\omega_g$ above the solution already found. There will be a simple orthogonal transformation to diagonalise $M^G(g)$ at this stationary point, and when it is applied with the g_i near that point, $M^G(g)$ is transformed to

$$M^G(g)' = k_g^{G2}\hbar\omega_g \begin{bmatrix} -3 & 0 & 0 & 0 \\ 0 & 1 & 0 & 0 \\ 0 & 0 & 1 & 0 \\ 0 & 0 & 0 & 1 \end{bmatrix} + k_g^G\omega_g\sqrt{\hbar\omega_g}$$

$$\times \begin{bmatrix} \frac{3}{2}(\epsilon_1 - \epsilon_3) & -\frac{1}{\sqrt{2}}\epsilon_2 & -\frac{1}{2}(\epsilon_1 + \epsilon_3) & -\frac{1}{\sqrt{2}}\epsilon_4 \\ -\frac{1}{\sqrt{2}}\epsilon_2 & \epsilon_3 & \frac{1}{\sqrt{2}}\epsilon_2 - \sqrt{2}\epsilon_4 & -\epsilon_1 - \epsilon_3 \\ -\frac{1}{2}(\epsilon_1 + \epsilon_3) & \frac{1}{\sqrt{2}}\epsilon_2 - \sqrt{2}\epsilon_4 & -\frac{1}{2}(\epsilon_1 - \epsilon_3) & -\sqrt{2}\epsilon_2 - \frac{1}{\sqrt{2}}\epsilon_4 \\ -\frac{1}{\sqrt{2}}\epsilon_4 & -\epsilon_1 - \epsilon_3 & -\sqrt{2}\epsilon_2 - \frac{1}{\sqrt{2}}\epsilon_4 & -\epsilon_1 \end{bmatrix}, \tag{4.13}$$

where ϵ_1 and ϵ_3 are small changes in g_1 and g_3 from the turning point. That is, $g_1 = -g_0 + \epsilon_1$ and $g_3 = g_0 + \epsilon_3$; and $g_2 = \epsilon_2$ and $g_4 = \epsilon_4$ are also small. Inspection of this form of the matrix shows that there is a second-order contribution to the lowest root of

$$\Delta E = -\frac{(k_g^G\omega_g\sqrt{\hbar\omega_g})^2 \frac{1}{2}(\epsilon_2^2 + \epsilon_4^2 + (\epsilon_1 + \epsilon_3)^2/2)}{4k_g^{G2}\hbar\omega_g}. \tag{4.14}$$

When the elastic energy $\frac{1}{2}\omega_g^2 \sum_i g_i^2$ is added, we get a total second-order contribution to the potential energy, which can be written

$$\Delta E = \frac{1}{2}\omega_g^2 \left(\frac{3}{4}(\epsilon_2^2 + \epsilon_4^2) + \frac{3}{8}(\epsilon_1 + \epsilon_3)^2 + \frac{1}{2}(\epsilon_1 - \epsilon_3)^2\right). \tag{4.15}$$

This shows that the stationary point is a minimum, because ΔE is positive definite. It also shows that the curvature of the APES at this point in the four-dimensional space is ω_g^2 in the $(g_1 - g_3)$ or g_0 direction and $\frac{3}{4}\omega_g^2$ in the three other orthogonal directions, so we should expect the spectrum of excitations of the system in this well at strong coupling to be that of a three-dimensional oscillator with frequency $\sqrt{\frac{3}{4}}\omega_g$ convoluted with that of a one-dimensional oscillator with frequency ω_g. This particular minimum gives rise to the T_h distortion whose picture is shown in Figure 2.6.

4.2.2 Biharmonic Parametrization of the G Bases

Before proceeding further with our discussion, it is useful to introduce a set of angular coordinates for use in a four-dimensional space, analogous to the better known spherical polar coordinates for three dimensions. The new choice can be defined in terms of cartesian coordinates in a four-dimensional g space as follows:

$$g_1 = g \sin\theta \sin\alpha, \quad g_2 = g \sin\theta \cos\alpha, \tag{4.16}$$
$$g_3 = g \cos\theta \sin\beta, \quad g_4 = g \cos\theta \cos\beta,$$

where $0 \leq g < \infty, 0 \leq \alpha < 2\pi, 0 \leq \beta < 2\pi$, and $0 \leq \theta < \pi/2$. They are normalized to g:

$$g_1^2 + g_2^2 + g_3^2 + g_4^2 = g^2. \tag{4.17}$$

In terms of these coordinates, the distance between two points in the space is given by

$$ds^2 = dg_1^2 + dg_2^2 + dg_3^2 + dg_4^2 = dg^2 + g^2 d\theta^2 + g^2 \sin^2\theta d\alpha^2 + g^2 \cos^2\theta d\beta^2, \tag{4.18}$$

and the volume element for integration in these coordinates is given by

$$dg_1 dg_2 dg_3 dg_4 = g^3 \cos\theta \sin\theta dg d\theta d\alpha d\beta. \tag{4.19}$$

In terms of these variables, the position of the minimum (4.11) can be written

$$\min 1 = (g, \theta, \alpha, \beta) = \left(\frac{3}{\sqrt{2}} k_g^G \sqrt{\frac{\hbar}{\omega_g}}, \frac{\pi}{4}, \frac{3\pi}{2}, \frac{\pi}{2} \right). \tag{4.20}$$

In a similar way the electronic bases can be written in terms of the biharmonic variables. As these bases are normalized, they only need the three angular parameters, and a general four-component basis can be written as

$$\mathbf{a} = (\sin\theta \sin\alpha, \sin\theta \cos\alpha, \cos\theta \sin\beta, \cos\theta \cos\beta). \tag{4.21}$$

In these terms the basis (4.10) is given by

$$(\theta, \alpha, \beta) = \left(\frac{\pi}{4}, \frac{3\pi}{2}, \frac{\pi}{2} \right). \tag{4.22}$$

There are four more minima on the lowest APES that are related to min 1 by rotations of the icosahedron, all having the same energy and the same value of g, which correspond to different T_h distortions of the icosahedron. They can be found by doing a numerical search for minima in the energy at the known value of g in the (θ, α, β) space. In terms of the biharmonic variables, the five minima are at $(g, \theta, \alpha, \beta)$ with the values of the variables shown in Table 4.1.

TABLE 4.1

Values of the $\{g_i\}$ at the $G \otimes g$ Minima in Terms of the Biharmonic Coordinates

	g	θ	α	β
min 1	g_0	$\pi/4$	$3\pi/2$	$\pi/2$
min 2	g_0	$\pi/4$	$7\pi/10$	$9\pi/10$
min 3	g_0	$\pi/4$	$3\pi/10$	$\pi/10$
min 4	g_0	$\pi/4$	$11\pi/10$	$17\pi/10$
min 5	g_0	$\pi/4$	$19\pi/10$	$13\pi/10$

, where $\quad g_0 = \dfrac{3}{\sqrt{2}} k_g^G \sqrt{\dfrac{\hbar}{\omega_g}}$

TABLE 4.2

The $\{g_i\}$ at the $G \otimes g$ Minima in Matrix Form

The actual values of the $\{g_i\}$ are given by multiplying the rows of Q_g^G by $k_g^G \sqrt{\frac{\hbar}{\omega_g}}$.

$$Q_g^G = \begin{bmatrix} -3/2 & 0 & 3/2 & 0 \\ p_3 & -p_1 & p_4 & -p_2 \\ p_3 & p_1 & p_4 & p_2 \\ -p_4 & -p_2 & -p_3 & p_1 \\ -p_4 & p_2 & -p_3 & -p_1 \end{bmatrix}, \quad \text{where} \quad \begin{cases} p_1 = (3/2)\cos(3\pi/10) \\ p_2 = (3/2)\cos(\pi/10) \\ p_3 = (3/2)\sin(3\pi/10) \\ p_4 = (3/2)\sin(\pi/10) \end{cases}$$

In terms of the g_i's the positions of the five minima are given by the rows of the matrix $k_g^G \sqrt{\frac{\hbar}{\omega_g}} \times Q_g^G$, where Q_g^G is given in Table 4.2. The electronic basis, $\mathbf{a}$, at each minimum is closely related to the $\{g_i\}$. The set of bases is given by the rows of the matrix $\mathbf{a}_g^G$, where

$$\mathbf{a}_g^G = \frac{\sqrt{2}}{3} Q_g^G. \tag{4.23}$$

The similarity of $\mathbf{a}_g^G$ and Q_g^G can be regarded as a result of the self-dual nature of $G \otimes g$. Both matrices represent the organisation of a set of bases for the G irrep under the same reduction of symmetry.

All the other stationary points in the energy that appear as solutions of the Öpik and Pryce equations are found not to be local minima when tested by second-order perturbation, so they do not represent possible static Jahn-Teller distortions.

Figure 4.1. A contour plot of the energy of the lowest APES in G ⊗ g plotted on an (α, β) surface, with α horizontal. The range of α and β is $(0, 4\pi)$, $\theta = \pi/4$, and g is constant. The minima are numbered to show how each one recurs with a period of 2π in each direction. The projections of the saddle points onto this surface are midway between neighbouring minima.

4.2.3 The Geometry of the Ground States

The situation we have reached here is similar to that discussed under the heading of *warping* in T ⊗ h in Chapter 3. As there, we have here a set of equivalent wells in a potential energy surface, and we assume that they are deep enough for the vibronic wave function to be concentrated in these wells. We further assume that the wave function is only very thinly spread in the region between them by tunnelling through the classically forbidden region. However, we can obtain a correct view of the classification of the low-lying energy states only by looking into what happens to the phase of the wave function in the tunnelling region. The geometry of these five minima can be simply displayed by looking at the surface containing them with g and θ constant (g_0 and $\pi/4$) and plotting α and β along cartesian axes. If also the range of values of (α, β) is extended so that a point at (α, β) reappears at $(\alpha + 2\pi, \beta)$ and so on, then these five minima form a square mesh, with every copy of one of them having as neighbors copies of each of the other four (Figure 4.1). On this plane the distances are just proportional to distances in the vibrational space, and our five minima are equidistant in both

TABLE 4.3

Strong Coupling Energies at Stationary Points of the
Systems $G \otimes g$ and $G \otimes h$

The † indicates solutions that are degeneracies for
one type of mode acting alone, though any interaction
with the other modes removes the degeneracy. The ‡
indicates a solution that is not on the lowest APES for
the one type of mode on its own.

Symmetry	Energy($G \otimes g$)	Energy($G \otimes h$)
$D_{3d}(I)$	$-\frac{3}{2}(k_g^G)^2 \hbar\omega_g$	$-\frac{3}{2}(k_h^G)^2 \hbar\omega_h$
$D_{2h}(II)$	$-\frac{1}{4}(k_g^G)^2 \hbar\omega_g$†	$-4(k_h^G)^2 \hbar\omega_h$
$D_{3d}(III)$	$-\frac{1}{6}(k_g^G)^2 \hbar\omega_g$‡	$-\frac{25}{6}(k_h^G)^2 \hbar\omega_h$
$T_h(IV)$	$-\frac{9}{4}(k_g^G)^2 \hbar\omega_g$	0†

spaces. We should remark here that five points equidistant from each other, and all the same distance from the origin, form the four-dimensional analogue of a tetrahedron. The next thing needed is to identify approximately a tunnelling path between neighboring mimina. The interaction between states in different wells will be dominated by tunnelling along the path of least resistance. That path will be a compromise between the shortest path and the one that goes through the lowest saddle point in energy between the minima. Accordingly we revisit the Öpik and Pryce equations in search of saddle points, and find ten of them, all related by symmetry. These saddle points are listed as type I in Appendix G (G.9) and in Table 4.3. Plotted on the (α, β) plane, each one lies at the midpoint of the line joining two adjacent minima, but at slightly different values of g and θ. We can thus expect the tunnelling path between adjacent minima to lie somewhere between the path of shortest distance, which is in the plane, and the (longer) path via the saddle point. This is enough information, as Berry's theorem (see Section 1.4) tells us that the phase changes along any two paths with the same endpoints are identical as long as a line of degeneracies does not come between these paths. A little further investigation locates the lines of degeneracy and verifies that they do not interpose between these two paths, so we can safely use the path of shortest distance to calculate the phase changes. We give the result of the phase tracking here, and the method in the next section. We restrict our attention to closed paths made up of straight-line segments between neighboring minima ("closed" paths begin and end on minima with the same number). The extended (α, β) space contains many such paths because it contains many copies of each minimum, and on any closed path that visits just one copy of each of the five minima, the sign of the electronic basis changes upon returning to a copy of the minimum from which it started. A closed path that visits a copy of every minimum twice returns the sign to its starting value. We have here then a similar result to that already found in T ⊗ h: two points in the vibrational phase space corresponding to the same *real* distortion have come out with opposite signs of the electronic basis because we have forced that basis to be real rather than single-valued. The single-valuedness of the whole vibronic wave function must then be restored by having a vibrational part that changes sign in step with the electronic basis. This can be achieved in this case by attaching signs to the copies of the minima in such a way that the sign alternates between pairs of neighbors. In this way the sign change round any loop of five will be properly represented. We then attach the same sign to the lowest vibrational wave function in each well, and look for states that are a linear combination of these product states in each well. The well-to-well overlaps will then be negative for all the neighboring wells, and because every minimum is a nearest neighbor of every other, the matrix of overlaps will be of the form

$$
S_g^G =
\begin{bmatrix}
0 & -S & -S & -S & -S \\
-S & 0 & -S & -S & -S \\
-S & -S & 0 & -S & -S \\
-S & -S & -S & 0 & -S \\
-S & -S & -S & -S & 0
\end{bmatrix}
\tag{4.24}
$$

with S a positive number. The diagonalization of the matrix S_g^G yields a G quartet state corresponding to an eigenvalue of $+S$ and an A singlet corresponding to an eigenvalue of $-4S$. The transformation that diagonalizes S_g^G then just reduces the space of the T_h minima in G ⊗ g to the irreducible spaces A and G of the icosahedral group. This transformation is closely related to the matrix Q_g^G (Table 4.2) in that the four normalized eigenvectors corresponding to the G eigenstates are the four columns of $2\sqrt{2}Q_g^G/(3\sqrt{5})$. The eigenvector for the A eigenstate is just $(1, 1, 1, 1, 1)/\sqrt{5}$. The vibronic wave functions are then found by taking linear combinations of ground state harmonic oscillator wave functions, using the eigenvectors as coefficients. For a fuller discussion of tunneling, see Appendix B. A consideration of the contribution of tunnelling to the energy, as described in Appendix B, shows that the positive overlap of the quartet state implies that it has a lower energy than that of the singlet of overlap $-4S$. The result is that the ground state after Jahn-Teller coupling has the same symmetry, G, as the original uncoupled electronic state.

4.2.4 Numerical Phase Tracking

An additional difficulty in G⊗g is that we lack an analytic form for the electronic basis at a general point, unlike the case for T ⊗ h. We are forced as a result to use a numerical method of phase tracking. Since the Jahn-Teller matrices are diagonalised numerically, the phases of the eigenvectors must also be tracked numerically. This is a little tricky because there is no intrinsic way of relating the phase of an eigenvector calculated at two neighboring points on the APES. One method easy to use is to travel along a path, calculating the eigenvector of the lowest eigenvalue at regular intervals and adjusting the overall sign of each eigenvector so that the first component is always positive. This produces continuity except where the first component would naturally go through zero, when all the other components are found to change sign with a very visible discontinuity. We then plot the components of the eigenvectors and count the number of discontinuities, which should of course occur simultaneously for all components. If this number is odd, then the eigenstate will have changed sign an odd number of times, implying an overall Berry phase change of π on completion of the adiabatic loop.

4.2.5 The Ham Factors in $G \otimes g$

We must remind ourselves that the Ham factors are the ratios of the matrix element of a particular operator within the ground set of Jahn-Teller states to the matrix element of the same operator within the original uncoupled basis. Here we need to look at any G operator, and the most available one to use is a g-type distortion. Only one component of the g set is needed, so for example we can start with the matrix $\mathbf{U}^G(G, 3)$, which gives the effect of a unit g_3 distortion in the original states.

$$\mathbf{U}^G(G, 3) = \begin{bmatrix} -1 & 0 & 1 & 0 \\ 0 & 1 & 0 & -1 \\ 1 & 0 & 0 & 0 \\ 0 & -1 & 0 & 0 \end{bmatrix} \tag{4.25}$$

Only one matrix element of (4.25) within one of the Jahn-Teller ground states is needed, so we look at the effect this distortion has on the basis vector $\sqrt{\frac{8}{45}}(0, -p_1, p_1, -p_2, p_2)$, which is the normalized second column of the matrix Q_g^G (Table 4.2), the second member, G_2 of the strongly coupled G quartet. If we describe the effect of $\mathbf{U}^G(G, 3)$ operating at the nth minimum as $\langle n|\mathbf{U}^G(G, 3)|n\rangle$, then the result is

$$\langle G_2|\mathbf{U}^G(G, 3)|G_2\rangle = \frac{8}{45}(p_1^2\langle 2|\mathbf{U}^G(G, 3)|2\rangle + p_1^2\langle 3|\mathbf{U}^G(G, 3)|3\rangle$$
$$+ p_2^2\langle 4|\mathbf{U}^G(G, 3)|4\rangle + p_2^2\langle 5|\mathbf{U}^G(G, 3)|5\rangle). \tag{4.26}$$

The quantities $\langle n|\mathbf{U}^G(G, 3)|n\rangle$ are the negative of the third column of the array (Table 4.2), $(3/2, p_4, p_4, -p_3, -p_3)$. This is because the five minima are solutions of the Öpik and Pryce equations, so that equation (4.8) holds. Finally

$$\langle G_2|\mathbf{U}^G(G, 3)|G_2\rangle = -\frac{8}{45}(p_1^2 p_4 + p_1^2 p_4 - p_2^2 p_3 - p_2^2 p_3) = 3/4, \quad (4.27)$$

and comparison with the 1 in the $(2, 2)$ position of $\mathbf{U}^G(G, 3)$ shows that the Ham factor in this case is 3/4. There is another aspect of Ham factors at strong coupling that must be mentioned here: Ham factors are a way of listing matrix elements of an operator within the vibronic ground states and so are of use when the operator only needs to be taken to first order in those states. Accordingly they give an incomplete picture when the operator couples the ground states with other states that are close in energy. In the case of $G \otimes g$, we have seen that there is an A state that can come very close to the G quartet when the coupling is strong, and an operator of G symmetry can connect A and G. Hence we need to know the strength of that coupling, and it can be worked out in exactly the same way as for the G Ham factor. The result can only be expressed in terms that relate it to the matrix of the G operator in the ground states. For instance the matrix of the g_3 operator can be written as

$$
\begin{array}{c|c|cccc}
 & A & G_1 & G_2 & G_3 & G_4 \\
\hline
A & 0 & 0 & 0 & -\dfrac{3}{2\sqrt{2}} & 0 \\
\hline
G_1 & 0 & -\dfrac{3}{4} & 0 & \dfrac{3}{4} & 0 \\
G_2 & 0 & 0 & \dfrac{3}{4} & 0 & -\dfrac{3}{4} \\
G_3 & -\dfrac{3}{2\sqrt{2}} & \dfrac{3}{4} & 0 & 0 & 0 \\
G_4 & 0 & 0 & -\dfrac{3}{4} & 0 & 0 \\
\end{array}
, \qquad (4.28)
$$

and at very strong coupling it is important not to forget these connecting terms. They cannot be defined as a ratio as can the normal Ham factors because their value is zero at zero coupling.

4.3 SYMMETRY AND THE TWO PHASE SPACES

The rather high symmetry, T_h, of the distortion of the icosahedron at the $G \otimes g$ minima can be regarded as a result of the *epikernel principle* propounded by Ceulemans and Vanquickenbourne (1989). This principle combines considerations of group theory, which show which symmetries can be produced by the activation of a particular set of normal coordinates, with a consideration of the energy gradient that shows a preference for distortions, or epikernals, of higher rather than lower symmetry. The principle is used by constructing sequences of successive permitted symmetry reductions, and stopping each sequence as soon as the electronic degeneracy is removed. At the end of each sequence is one of the preferred epikernels. The results of this process do not predict which of the expected distortions are actually minima, but it is very useful in providing a list of possibilities, and all the preferred epikernels do turn up as stationary points. The adiabatic approximation in which we are working focuses attention on the phase space of the vibrations, the $\{g_i\}$ here, because this is the space in which the Schrödinger equation must ultimately be solved. This is also the space in which the pseudo-rotational states exist. However, the Öpik and Pryce procedure works in the electronic basis, the $\{a_i\}$, and this basis can also be regarded as spanning a phase space. The lowest APES can be followed in both spaces, and the similarity of the bases at the minima, $\mathbf{a}_g^G$, to the $\{g_i\}$ at the minima (4.23) points to a close relationship between the two. This relationship can also be accounted for in terms of the epikernel principle. The descent in symmetry associated with the G bases follows through in an identical manner to that associated with the g coordinates: there are epikernels in the electronic space equivalent to those in vibrational space, so naturally the basis vectors arrived at in the same way are equivalent. A further advantage in looking at the epikernels in the electronic basis vector space is that we can expect to find stationary points at the same points in this space for different sets of vibrations, as will be seen later.

4.4 $G \otimes h$

The description of this case can be made quite brief because it is very similar to $G \otimes g$, and because of the work done on the h modes in Chapter 3. The Hamiltonian can be written similarly to that for $G \otimes g$ as

$$\mathcal{H} = -\frac{1}{2}\hbar^2 \sum_{i=1}^{5} \frac{\partial^2}{\partial h_i^2} + \frac{1}{2}\omega_h^2 \sum_{i=1}^{5} h_i^2 + M^G(h), \qquad (4.29)$$

where the interaction matrix, listed in Appendix E (E.14), is

$$M^G(h) = k_h^G \omega_h \sqrt{\hbar\omega_h} \qquad (4.30)$$

$$\times \begin{bmatrix} 2h_2 - \sqrt{3}h_1 & 2h_5 & h_2 - h_4 & h_3 - h_5 \\ 2h_5 & -\sqrt{3}h_1 - 2h_2 & -h_3 - h_5 & -h_2 - h_4 \\ h_2 - h_4 & -h_3 - h_5 & \sqrt{3}h_1 - 2h_4 & -2h_3 \\ h_3 - h_5 & -h_2 - h_4 & -2h_3 & \sqrt{3}h_1 + 2h_4 \end{bmatrix}$$

and the $\{h_i\}$ are the $\{Q_i\}$ normal coodinates of Chapter 3. All the methods described for $G \otimes g$ can be used in a search for the minima and other stationary points in this case. As before, some solutions of the Öpik and Pryce equations can be found with two components of the electronic base equal to zero. These provide a useful starting point from which to generate other solutions. The minima on the lowest APES turn out to be at distortions that have already appeared in Chapter 3 as one of the possible sets of minima of the warped $T \otimes h$ system. These are the ten D_{3d} points on the spherical subspace $\alpha = 0$ shown in Figure 3.2 (using the $[Q, \alpha, \gamma, \theta, \phi]$ parametrization for the $\{h_i\}$, equation [3.5])[2]. In view of the epikernel principle, this is not a surprising coincidence. The six D_{5d} points on the same spherical subspace are lines of degeneracy in the lowest APES of $G \otimes h$ rather than stationary points. The set of fifteen D_{2h} points that are saddle points in Figure 3.2 are not stationary points here, but $G \otimes h$ does have a corresponding set of fifteen D_{2h} saddle points lying just off the $\alpha = 0$ surface. This set of saddle points lie very nearly on the shortest path between pairs of minima, and also provide the lowest energy barrier between them, so the ground state vibronic wave function will be concentrated near the $\alpha = 0$ surface. The energy at the minima is $-\frac{25}{6} k_h^{G^2} \hbar \omega_h$, and at the saddle points it is $-4 k_h^{G^2} \hbar \omega_h$. These minima are labelled as Type III and the saddle points as Type II in Table 4.3 and in Appendix G, where the basis and vibrational coordinate vectors are tabulated.

4.4.1 Phase Tracking and the Ground States

The D_{3d} points on the spherical subspace are situated at the vertices of a dodec-ahedron on an $\alpha = 0$ spherical subspace, and pairs of vertices that are related by inversion of the dodecahedron correspond to the same distortion, or the same set of values of the $\{h_i\}$. Thus each of the ten minima corresponds to two of the twenty vertices. The saddle points are close to the center of every edge, and again there are two copies of each one. As before, numerical phase tracking can be used to follow the sign of the electronic wave function over the APES, and the paths tracked can be restricted to those following the shortest distance from minimum to minimum on the $\alpha = 0$ surface, which are paths that map onto an edge of the dodecahedron. It is found that the wave function changes sign along any path made up of a sequence of five edges that begins and ends on a particular minimum. Two such paths are shown in Figure 4.2, one open and one closed on the dodecahedron. Both of them contain an odd number of links, so if it is assumed that the phase changes sign once on each edge, then the sign of the wave function will always be correctly given. This situation mirrors closely that in $G \otimes g$. As in that case, a matrix of overlaps is set up with $-S$ joining every pair of minima that are neighbors on the dodecahedron. As each

---

[2]In this section, α is the parameter that appears in the parametrization of the h modes, (3.5), not the α of the biharmonic parametrization of the g modes.

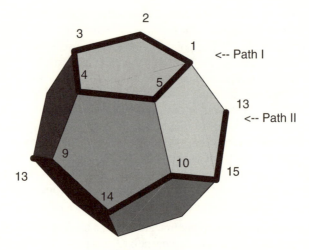

Figure 4.2. The mapping of the G $\otimes$ h minima on a dodecahedron. Both the paths shown induce a change of sign on the electronic wave function.

minimum has three others as nearest neighbors, this matrix takes the form

$$
S_h^G =
\begin{bmatrix}
0 & -S & 0 & 0 & -S & -S & 0 & 0 & 0 & 0 \\
-S & 0 & -S & 0 & 0 & 0 & -S & 0 & 0 & 0 \\
0 & -S & 0 & -S & 0 & 0 & 0 & -S & 0 & 0 \\
0 & 0 & -S & 0 & -S & 0 & 0 & 0 & -S & 0 \\
-S & 0 & 0 & -S & 0 & 0 & 0 & 0 & 0 & -S \\
-S & 0 & 0 & 0 & 0 & 0 & 0 & -S & -S & 0 \\
0 & -S & 0 & 0 & 0 & 0 & 0 & 0 & -S & -S \\
0 & 0 & -S & 0 & 0 & -S & 0 & 0 & 0 & -S \\
0 & 0 & 0 & -S & 0 & -S & -S & 0 & 0 & 0 \\
0 & 0 & 0 & 0 & -S & 0 & -S & -S & 0 & 0
\end{bmatrix}.
\tag{4.31}
$$

Diagonalization yields G, H, and A states, with overlaps $2S$, $-S$, and $-3S$ respectively. The G states, having the maximum overlap, have the minimum energy. (see Appendix B)

The eigenvectors of this matrix can be quoted in terms of the matrices listed in Appendix G: the (normalized) eigenvectors corresponding to the four G states are the four columns of $\sqrt{3/50} \times \mathbf{a}_h^G$ (Table G.7), the (normalized) eigenvectors corresponding to the five H states are the five columns of $\sqrt{2/5} \times Q_h^G$ (Table G.5), while the eigenvector corresponding to the A state is a vector with $1/\sqrt{10}$ in each position.

4.4.2 The Ham Factors in $G \otimes h$

The Ham factors are produced in the same way as is described in Section 4.2.5, but now that we have a matrix (4.30) from which matrices of H-type operators in the G states such as $U^G(H, 1)$ can be derived, we can find the Ham factors for H as well as G operators. Carrying through the procedure of Section 4.2.5 gives the following strong-coupling results:

$$
\begin{array}{c|cc}
 & K(G) & K(H) \\
\hline
G \otimes g & \frac{3}{4} & 0 \\
G \otimes h & \frac{5}{90} & \frac{5}{9}
\end{array}
\qquad (4.32)
$$

This table illustrates the phenomena of quenching. The fact that $K(H) = 0$ in $G \otimes g$ at strong coupling means that a small perturbation of h symmetry, such as a stress, will have no effect on the ground states in that case. The Jahn-Teller interaction with the h modes is quenched by the interaction with the g modes. In a similar way the g mode interaction is almost quenched if the interaction with the h modes is strong. It is also the case that $K(T_1) = 0$ in this strong coupling regime, because all the electronic bases are real on the lowest APES, so that we can also say that the angular momentum, which is a T_1 operator, is completely quenched by the Jahn-Teller interaction at strong coupling. The question of off-diagonal matrix elements mentioned in Section 4.2.5 becomes quite complicated here because of the number of ground states. We shall not go into it further here, but refer the reader to Cullerne et al. (1995).

4.5 $G \otimes (g \oplus h)$

This problem with nine different vibrational coordinates looks much more complicated than what has gone before, but thanks to the method of Öpik and Pryce with its equations in terms of the four electronic bases, there are in fact few extra difficulties. Inspection of the equations in Section 4.2.1 shows that a given basis vector, **a**, can only be at a stationary point for the joint set of vibrational coordinates if it is at a stationary point for each set separately. The result is that the set of four different types of stationary points already found is the same set that must be considered here. The complete list of energies is given in Table 4.3. If both types of mode are active, the energy at a stationary point is just the sum of the energies in the two columns. This is illustrated in Figure 4.3. From this plot it is very obvious that only three situations arise. Whenever the g-coupling predominates, the minima are of Type IV and the path of lowest energy between them is via the Type I saddle points. Conversely if the h-coupling is strongest, the minima are of Type III and the intermediate saddle points are of Type II.

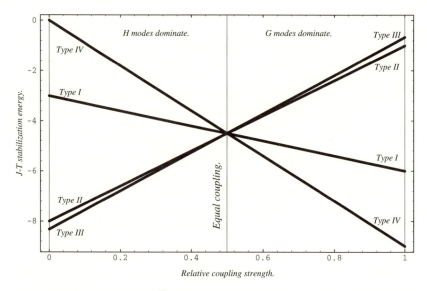

Figure 4.3. The energies of the Öpik and Pryce stationary points in $G \otimes (g \oplus h)$, showing how they vary with relative coupling strength

The relative coupling strength at which all the above points coincide in energy is given by $(k_g^G)^2 \hbar\omega_g = 2(k_h^G)^2 \hbar\omega_h$. This is called the equal coupling regime, $G \otimes (g \oplus h)_{eq}$, and will be discussed shortly. To discuss the complete structure of degeneracies and Berry phases over the full nine-dimensional phase space of $G \otimes (g \oplus h)$ would be a mammoth task. Indeed, it is a task that we did not even complete in the four- and five-dimensional phase spaces of the subsystems $G \otimes g$ and $G \otimes h$. What we did do and what is needed now is to understand how the phases change over the minimum energy tunneling paths between minima. The energy plot in Figure 4.3 suggests that at all relative coupling strengths the minima and the minimum energy paths connecting the minima are either the same or closely related to those calculated for either the g or the h mode regimes separately. This means that as we move from coupling to one phonon type alone and increase the influence of the other phonon type, the topology of the minima, the tunneling paths that connect them, the phases, and the ground states will be constructed in just the same way as before. Take for example the case when the g mode dominates. Based on our experience with $G \otimes g$ in Section 4.2, we know that there will be five energy minima corresponding to tetrahedral (T_h) distortions. These five minima are of type (IV), and the minimum energy paths available for tunneling are via the type (I) saddles. The difference now is that the coupling to the h modes reduces the energy of the saddles but increases the length of the minimum energy tunneling paths. The path that dominates the tunneling energy will be a compromise between these two effects, but as long as the old path can be continuously distorted into the new one without crossing a degeneracy on the lowest APES, the phase relationships will be preserved in following through from $G \otimes g$ to $G \otimes (g \oplus h)_{eq}$, preserving the G-type ground state. In a similar vein we can consider what happens as one approaches $G \otimes (g \oplus h)_{eq}$ from $G \otimes h$. Referring to Section 4.4 we can recall that the ten energy minima correspond to D_{3d} distortions, and that there are fifteen saddle points between minima at points of D_{2h} symmetry near the $\alpha = 0$ surface. In this case we check that as we increase the strength of g distortions on the $G \otimes h$ subsystem, the points corresponding to D_{3h} symmetry remain mutually equidistant. As we approach $G \otimes (g \oplus h)_{eq}$ but with the h modes still dominant, the tunneling between the D_{3h} points still occurs via the type (II) saddles. The coupling to the g modes reduces the energy of these saddles but increases the length of the minimum energy tunneling paths. The form of the overlap matrix between the D_{3h} points remains unchanged until we reach the equal coupling regime, so we carry the argument for identifying the vibronic ground states through from $G \otimes h$ coupling to equal $G \otimes (g \oplus h)_{eq}$ coupling, and conclude that the G-type ground state is preserved for this range of parameters too.

4.5.1 $G \otimes (g \oplus h)_{eq}$ and SO(4) Symmetry

At the beginning of this chapter, we remarked that there was one choice of relative coupling strengths at which the Hamiltonian is invariant under the operations of the symmetry group SO(4), and this choice must clearly be identified with the choice that makes all the stationary points of equal energy. The existence of the SO(4) invariance shows that the energy must be constant over a four-dimensional subspace of the full nine-dimensional coordinate space. We thus expect to find the minimum energy over a four-dimensional trough exactly as the minimum energy in $T \otimes h$ is constant over a three-dimensional trough. The similarity extends to the fact that, whereas in $T \otimes h$ we have an $L = 1$ representation for the electronic state and an $L = 2$ representation for the vibrations, here we have a $[1, 0]$ representation for the electronic state and a $[2, 0]$ representation for the vibrations (using SO(4) labels). To get the SO(4) invariance of the whole Hamiltonian, including the kinetic energy term, it is necessary to have $\omega_g = \omega_h$ in addition to the condition that flattens the potential energy surface, and this assumption will be made now. Any difference in the frequencies can be introduced as a perturbation, along with any warping of the energy surface. The method adopted to discuss this situation is to use a set of coordinates natural to the SO(4) group: the biharmonic coordinates already introduced in Section 4.2.2. There they were used to represent the electronic states, the four components of the $[1, 0]$ representation. The nine vibrational coordinates associated with the nine components of a $[2, 0]$ representation can be written as

$$q_1 = \frac{q}{\sqrt{3}} \cos 2\theta, \tag{4.33}$$

$$q_2 = \frac{q}{\sqrt{3}} \sin 2\theta \cos(\alpha + \beta), \quad q_3 = \frac{q}{\sqrt{3}} \sin 2\theta \sin(\alpha + \beta),$$

$$q_4 = \frac{q}{\sqrt{3}} \sin 2\theta \cos(\alpha - \beta), \quad q_5 = \frac{q}{\sqrt{3}} \sin 2\theta \sin(\alpha - \beta),$$

$$q_6 = q\sqrt{\frac{2}{3}} \sin^2 \theta \cos 2\alpha, \quad q_7 = q\sqrt{\frac{2}{3}} \sin^2 \theta \sin 2\alpha,$$

$$q_8 = q\sqrt{\frac{2}{3}} \cos^2 \theta \cos 2\beta, \quad q_9 = q\sqrt{\frac{2}{3}} \cos^2 \theta \sin 2\beta.$$

The nine $\{q_i\}$ defined here are related by a linear transformation to the set of four $\{g_i\} \oplus$ five $\{h_i\}$, but rather than give the transformation explicitly we shall give the Jahn-Teller Hamiltonian in terms of the new coordinates, but in the same electronic basis states:

$$\mathcal{H} = -\frac{1}{2} \sum_{i=1}^{9} \frac{\partial^2}{\partial q_i^2} + \frac{1}{2} \sum_{i=1}^{9} q_i^2 + M^G(eq), \tag{4.34}$$

with

$$M^G(eq) = k_{eq}^G \begin{bmatrix} q_1 + \sqrt{2}q_6 & -\sqrt{2}q_7 & q_2 - q_4 & -q_3 - q_5 \\ -\sqrt{2}q_7 & q_1 - \sqrt{2}q_6 & -q_3 + q_5 & -q_2 - q_4 \\ q_2 - q_4 & -q_3 + q_5 & -q_1 + \sqrt{2}q_8 & -\sqrt{2}q_9 \\ -q_3 - q_5 & -q_2 - q_4 & -\sqrt{2}q_9 & -q_1 - \sqrt{2}q_8 \end{bmatrix}. \tag{4.35}$$

Because of the condition $\omega_g = \omega_h = \omega_{eq}$, we have been able to give the Hamiltonian its simpler form with $\hbar$ and ω_{eq} omitted and energies measured in units of $\hbar\omega_{eq}$ as in Chapter 3. As expected, the eigenvalues of $M^G(eq)$ are independent of the angles, the lowest root is at $-\sqrt{3}k_{eq}^G q$, and the other three roots are degenerate at $1/\sqrt{3}k_{eq}^G q$. Hence the minimum energy trough occurs at $q = \sqrt{3}k_{eq}^G$ at an energy $-3/2(k_{eq}^G)^2$. The eigenstate corresponding to the lowest energy is

$$|u\rangle = \sin\theta \sin\alpha |1\rangle + \sin\theta \cos\alpha |2\rangle + \cos\theta \sin\beta |3\rangle + \cos\theta \cos\beta |4\rangle, \tag{4.36}$$

where the kets $|i\rangle$, $i = 1, 2, 3, 4$ are the G electronic bases. Next we can look for the pseudo-rotational states on this trough, for which a form of the kinetic energy is needed. From (4.33) we can find

$$\sum_{i=1}^{9} dq_i^2 = dq^2 + \frac{8}{3}q^2(d\theta^2 + \sin^2\theta d\alpha^2 + \cos^2\theta d\beta^2). \tag{4.37}$$

Because the parametrization of the nine $\{q_i\}$ in terms of q and three angles is incomplete, we cannot get a full form for the Laplacian, but the terms involving motion along the minimum energy surface can be deduced from (4.37) to be

$$\nabla^2_{eq} = \frac{3}{8q^2} \left(\frac{1}{\sin\theta\cos\theta} \frac{\partial}{\partial\theta} \sin\theta\cos\theta \frac{\partial}{\partial\theta} + \frac{1}{\sin^2\theta} \frac{\partial^2}{\partial\alpha^2} + \frac{1}{\cos^2\theta} \frac{\partial^2}{\partial\beta^2} \right),$$
(4.38)

evaluated at $q^2 = 3(k^G_{eq})^2$. The Schrödinger equation for motion on the minimum energy surface can thus be written

$$-\frac{1}{2}\nabla^2_{eq}\psi - \frac{3}{2}(k^G_{eq})^2\psi = E\psi.$$
(4.39)

The properties and use of biharmonic coordinate systems are developed in very general terms by Barut and Rączka (1986), pp. 303–307, and the results needed here can be extracted from their work. The permitted solutions of (4.39) give

$$E(\ell) = -\frac{3}{2}(k^G_{eq})^2 + \frac{1}{16(k^G_{eq})^2}\ell(\ell+2) \quad \text{where} \quad \ell = 0, 1, 2, \dots. \quad (4.40)$$

The degeneracy of the states at $E(\ell)$ is $(\ell + 1)^2$, and they form a basis for the $[\ell, 0]$ representation. The eigenfunctions take the form

$$\psi = d^{(j)}_{\mu,\mu'}(2\theta)e^{im_1\alpha}e^{im_2\beta}$$
(4.41)

where $d^{(j)}_{\mu,\mu'}$ is the same function as that which appears in the rotation matrix elements in SO(3), as detailed in Chapter 4 of Edmonds (1960). The index j is positive and either integral or half-integral, and μ and μ' run from $-j$ to j in integral steps. The indices j, μ and μ' are related to ℓ, m_1, and m_2 by

$$\ell = 2j, \quad m_1 = \mu - \mu', \quad m_2 = \mu + \mu'.$$
(4.42)

Thus m_1 and m_2 are even or odd according as ℓ is even or odd. Examples of the use of (4.41) and (4.42) are the biharmonic parametrization of a g base (4.16), with $\ell = 1$, and the $\{q_i\}$ in (4.35), with $\ell = 2$. As usual we must end by taking account of the phases. Under the transformation

$$(\theta, \alpha, \beta) \to (\theta, \alpha + \pi, \beta + \pi),$$
(4.43)

the electronic basis changes sign, but as the $\{q_i\}$ are unchanged, this transformation must take us to the same point in coordinate space. Consequently the vibrational wave function $\psi(Q)$ must change sign to cancel the change of sign of the electronic wave function $|u\rangle$ in order that the total wavefunction $\Psi = \psi|u\rangle$ be single-valued. This means that the permitted values of ℓ must be odd integers. The energies are thus those given in (4.40), but with $\ell = 1, 3, \dots,$

and the lowest state with $\ell = 1$ is a $[1, 0]$ representation, or a vibronic quartet of symmetry G. The energies quoted in (4.40) omit certain terms of order $1/(k_{eq}^G)^2$ that do not depend on ℓ. These corrections were found for $T \otimes h$, but to find terms of this order correctly here, we would need to have a complete parametrization of the nine coordinates. The energies also omit the zero-point energy of the oscillators.

4.5.2 The Ham Factors

The four different Ham factors found for G and H operators in $G \otimes g$ and $G \otimes h$ now reduce to one, that for the $[2, 0]$ representation in $G \otimes (g \oplus h)_{eq}$ (this follows from the fact that the irreps G and H are both contained in the SO[4] irrep $[2, 0]$). This Ham factor can be found in the form of a ratio of integrals by considering the operation of a Q_1 operator in one component of the $[1, 0]$ vibronic state in exactly the same way as was described for $T \otimes h$ in Chapter 3, Section 3.3.2. The result is 1/3, so the table of Ham factors (4.32) can be filled in:

	$K(G)$	$K(H)$
$G \otimes g$	$\frac{3}{4}$	0
$G \otimes (g \oplus h)_{eq}$	$\frac{1}{3}$	$\frac{1}{3}$
$G \otimes h$	$\frac{5}{90}$	$\frac{5}{9}$

(4.44)

In each column the value for equal coupling lies between the values for coupling to one mode only. It is possible to show by the numerical work described in Section 4.5.3 that the values for other relative coupling strengths vary smoothly across the range; the quenching of one mode by the other gradually disappears as the relative coupling strengths change.

4.5.3 Other Relative Coupling Strengths

It is possible to extend the calculations outwards on both sides of the case $G \otimes (g \oplus h)_{eq}$ by introducing a difference in the coupling strengths, or equivalently, a difference of frequencies, as a perturbation on the pseudo-rotational states (4.41). This perturbation can be modelled as a warping potential that is an icosahedral invariant with minima either at the T_d points and maxima at D_{3d}, or vice versa. The simplest form of such a potential is given by Cullerne (1995). It is a fourth-order scalar in the $\{g_i\}$. On the $G \otimes (g \oplus h)_{eq}$ minimum energy surface, it takes the form

$$V_4 = (1 + 3 \cos 4\theta) - 16 \cos(\alpha - 3\beta) \cos^3 \theta \sin \theta + 16 \cos(3\alpha + \beta) \cos \theta \sin^3 \theta.$$
(4.45)

Because the degeneracies of the pseudo-rotational states increase so rapidly with energy, this matrix becomes very large, but it is possible to follow the states to large enough values of $\pm V_4$ to see the energy level structure moving significantly towards that of the unequal coupling regimes, and to see the Ham factors moving smoothly to the values appropriate to coupling to either g or h modes alone.

4.6 BROAD BAND SPECTRA

As remarked in the previous section, there is as yet no way of getting at energy levels for intermediate strengths of coupling, so the corresponding band shapes cannot be calculated. However it is possible to calculate band shapes in the strong coupling limit, using the Franck-Condon principle in a method attributed to Cho (1957). Starting from (3.74), which relates the transition probability to the overlap of vibronic functions, we see that the intensity distribution of a band arising from a set of transitions from a ground state ψ_0 to all the excited states ψ_i can be written as

$$I(E) = \sum_i |\langle \psi_0(\mathbf{Q}) | \psi_i(\mathbf{Q}) \rangle|^2 \delta(E_i - E_0 - E), \qquad (4.46)$$

where $\mathbf{Q}$ is a displacement vector in the space of the vibrational coordinates. In the limit of strong coupling, this sum has to be replaced by an integral over a continuum of states. In this limit, while the starting state is the well-defined lowest energy level of a harmonic oscillator, the finishing states are all at large energies above the minima of their APES wells and thus have high quantum numbers. At very high quantum numbers, the wave function of a bound state becomes concentrated at those points in coordinate space, $\mathbf{Q}_i$, where the total energy of the state is just equal to the potential energy, so the overlap required is just proportional to the amplitude of ψ_0 evaluated at that point, namely, $\psi_0(\mathbf{Q})\delta(\mathbf{Q} - \mathbf{Q}_i)$. Here we assume that the ground state is the lowest harmonic oscillator state, which is a Gaussian function in all the coordinates, centered at the origin:

$$\psi_0(\mathbf{Q}) = \exp\left(-\frac{1}{2} \sum_j Q_j^2\right), \qquad (4.47)$$

where j numbers the different coordinates, and redundant constants have been scaled out. For these Jahn-Teller systems a sum must be taken over the several APESs as well as the integral over coordinate space, and the final expression for the intensity of the transition is

$$I(E) = \sum_n \int_{\mathbf{Q}} \exp\left(-\sum_j Q_j^2\right) \delta(V_n(\mathbf{Q}) - E) d\mathbf{Q}, \qquad (4.48)$$

where $V_n(\mathbf{Q})$ is the energy of the nth APES at $\mathbf{Q}$. The integrand represents "vertical" transitions because the δ function picks out the energy $V_n(\mathbf{Q})$ of the point that is directly above the point $\mathbf{Q}$ in the ground state, assuming that energy is plotted vertically. The method of evaluation devised by Cho follows by summing up a large number of contributions to $I(E)$ produced by evaluating the integrand at a random set of points in coordinate space. The process uses a random number generator that produces a set of numbers with a Gaussian distribution, so that the exponential term in the integrand is handled. For every chosen point $\mathbf{Q}$, the energies of the APESs are worked out, and a histogram of all the energies found in this way gives the band shape. Some results for G electronic states calculated in this way by Cullerne (1995) are shown in Figures 4.4–4.7. The coordinate space is either that of the g or h modes or both together with a single a_g mode, that is, a 4-, 5-, or 10-dimensional coordinate space. These calculations also include the effect of spin-orbit coupling, which will in general split a ^2G state into states labelled Γ_7 and Γ_9 irreps of the double group, a doublet and a sextet state (see Table 6.1 for irreps of states with spin). There are always just four APESs to be considered, as the Jahn-Teller coupling does not remove the degeneracy of Kramers-conjugate pairs of states. All the band shapes have been scaled to the same half width and their centers aligned. The numbers on the vertical axes show the sample size per bin of the histogram. These spectra correspond to a physical situation where the ratio of the Jahn-Teller coupling strength to $\hbar\omega$ is large.

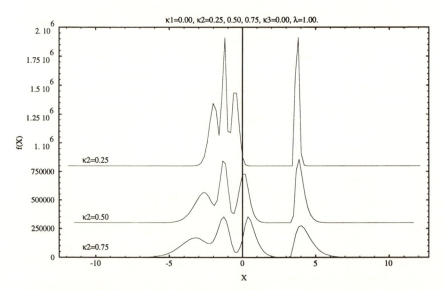

Figure 4.4. Band shapes with spin-orbit coupling (λ) and smaller amounts of h-mode coupling (whose strength is proportional to κ_2). Note the transition from the separate Γ_7 and Γ_9 states to four more equally spaced APESs.

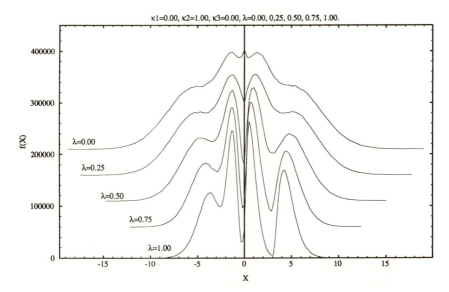

Figure 4.5. Band shapes with *h* mode coupling of strength κ_2 and various amounts of spin-orbit coupling (λ). κ_2 and λ are given in units of $\hbar\omega$. The strange extra peak at the center of the $\lambda = 0$ band does seem to be really there.

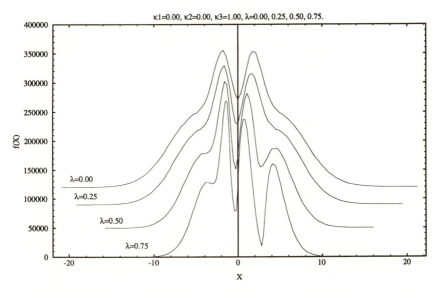

Figure 4.6. Band shapes with predominantly *g* mode coupling (κ_3) and various amounts of spin-orbit coupling (λ). κ_3 and λ are given in units of $\hbar\omega$.

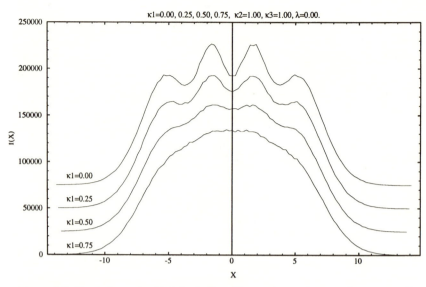

Figure 4.7. Band shapes for $G \otimes (g \oplus h)_{eq}$. Here an increasing coupling to a totally symmetric a_g mode (coupling strength proportional to κ_1) is included. This shows the effect of symmetric mode coupling in smoothing out structure.

4.7 AN OVERVIEW OF $G \otimes (g \oplus h)$

It is the size of this problem that prevents us from getting detailed solutions for any but extreme values of the parameters. The strong coupling regime has been most fully studied, though even there we have had to rely on a combination of perturbation and interpolation techniques between the single mode and equal mode regimes. Weak coupling can be handled by perturbation theory, which is straightforward but tedious to do and produces no surprises. For intermediate strengths of equal coupling, we might hope to be able to use the SO(4) symmetry in a similar way to the use of the SO(5)⊃SO(3) chain in $T \otimes h$, but this has not yet proved possible. However we can still use $T \otimes h$ and other simpler systems as models for interpolation, and Figure 3.5 shows the type of result to be expected for the Ham factors. Energy level plots like Figures 3.3 and 3.4 can also be sketched by doing a symmetry analysis of the states at strong and weak coupling and using perturbation at weak coupling. Energy levels and Ham factors between them comprise most of what could be measured in principle in the ground state. Thus it would appear that we can get a good idea of how the solutions behave in the unknown interior by approaching this problem from around all the edges.

5

Electronic Quintets and $H \otimes (g \oplus h)$

5.1 INTRODUCTION

The H irrep is the highest-dimensioned irreducible representation for the icosahedral group (I_h), and quintet electronic states are the most energetically degenerate that we can analyze without including spin. Quintet electronic states arise in several physical contexts—principally in cations of C_{60} with their H_u states, though the electronic ground states of icosahedral clusters such as Si_{12} also have been suggested as possible examples (Gu et al., 1993). The earliest theoretical attempt to analyze the Jahn-Teller modes for a quintet system was carried out by Khlopin, Polinger, and Bersuker, who discussed in particular the $H \otimes h$ system in 1978. Judd (1984) and Pooler (1978, 1980) provided further insights into the symmetries of quintet-based systems, using group theoretical arguments, and Ceulemans and Fowler (1990) have derived the positions of the extremal points on the adiabatic energy surfaces for the general $H \otimes (g \oplus h)$ interaction.

The symmetric part of the Kroneker product $H \otimes H$ is fifteen-dimensional and contains G and A once and H twice. This double appearance of H means that the Wigner-Eckart theorem does not take its usual simple form, and there are two families of matrices that must be multiplied by independent coupling constants to represent a general Jahn-Teller interaction. The precise choice of these matrices is free, insofar as once any pair has been found, any pair of linear combinations of them will do equally well. One choice is to take the matrices that correspond to the two H operators that derive either from a set of $L = 2$ states or from a set of $L = 4$ states. We take them operating within an $L = 2$ basis to produce the matrices $M_2^H(h)$ and $M_4^H(h)$ given in Appendix E (E.16, E.17). Another choice of two different H operators, corresponding to the one made by Ceulemans and Fowler (1990), taken within an $L = 2$ basis gives rise to the matrices $M_a^H(h)$ and $M_b^H(h)$ (E.18, E.19). The two sets of matrices are related by

$$M_a^H(h) = \frac{3\sqrt{3}}{7} M_2^H(h) - \frac{1}{7} M_4^H(h) \tag{5.1}$$

$$M_b^H(h) = -\frac{\sqrt{3}}{7} M_2^H(h) + \frac{1}{21} M_4^H(h)$$

In this chapter both choices ($M_2^H[h]$ and $M_4^H[h]$; $M_a^H[h]$ and $M_b^H[h]$) will be used where appropriate. The Jahn-Teller interaction with any actual h-type normal coordinate will correspond to some linear combination of the matrices from either set.

Pooler (1980) has examined the various coupling cases and regimes that arise with an H electronic state, with an emphasis on determining which special cases lead to additional symmetry in the Hamiltonian. He found the following possibilities, each of which possesses a particular set of coupling parameters leading to extra symmetry:

$$
\begin{aligned}
&(1) \quad \text{H} \otimes (g \oplus h_2 \oplus h_4)_{eq} : \quad \text{SO(5) symmetry;} \\
&(2) \quad \text{H} \otimes (g \oplus h_4)_{eq} : \quad \text{SO(3) symmetry;} \\
&(3) \quad \text{H} \otimes h_2 : \quad \text{SO(3) symmetry.}
\end{aligned}
\tag{5.2}
$$

Here h_2 indicates a mode whose interaction is given by an $L = 2$ type matrix, and h_4 is a mode with an $L = 4$ type matrix. In case (1) there are two different h modes, one of each type, so the interaction is with a fourteenfold set of vibrational modes. In case (2) there are nine vibrational modes, and in case (3), five modes. The lowest APES of each of these systems has SO(5) symmetry, which is surprising given the SO(3) symmetries of the Hamiltonians in cases (2) and (3).

An interesting result of the double appearance of the h operator is that the Ham factor $K(\text{H})$ in an H vibronic state will no longer be a single constant. Operating with, for example, an operator that produces an h_2 matrix in the uncoupled state might produce a matrix that is a linear combination of h_2 and h_4 matrices: in general, $K(\text{H})$ is a 2×2 matrix. This has been set out by Culierne et al. (1995), and we shall report their results under the appropriate headings.

The fullest account of H $\otimes$ ($g \oplus h$) published so far is that of Ceulemans and Fowler (1990) (see Section 5.8). Their analysis covers the geometry and topology of the lowest APES for all relative coupling strengths and different H $\otimes$ h matrices.

5.2 H $\otimes h_4$

We work from the particular to the general, and start the detailed discussion with the special case that corresponds to an H electronic state coupled to a single h mode through the interaction matrix, $M_4^H(h)$.

$$M_4^H(h) = k_{h,4}^H \times$$
$$\begin{bmatrix} 6\sqrt{3}h_1 & -4\sqrt{3}h_2 & \sqrt{3}h_3 & \sqrt{3}h_4 & -4\sqrt{3}h_5 \\ -4\sqrt{3}h_2 & -4\sqrt{3}h_1+2h_4 & -7h_3+2h_5 & 2h_2+7h_4 & 2h_3 \\ \sqrt{3}h_3 & -7h_3+2h_5 & \sqrt{3}h_1-7h_2 & -7h_5 & 2h_2-7h_4 \\ \sqrt{3}h_4 & 2h_2+7h_4 & -7h_5 & \sqrt{3}h_1+7h_2 & -7h_3-2h_5 \\ -4\sqrt{3}h_5 & 2h_3 & 2h_2-7h_4 & -7h_3-2h_5 & -4\sqrt{3}h_1-2h_4 \end{bmatrix} \quad (5.3)$$

It would be a heavy task to solve for the minima and other stationary points from scratch, and it is with relief that we note that the answers are already at hand. To see this, recall that the study of the system $G \otimes h$ in Chapter 4 provides values of sets of normal coordinates $\{h_i\}$ corresponding to the allowed Jahn-Teller distortions to D_{3d} or D_{2h} symmetry. These are listed in Appendix G in Tables G.5 and G.11. Now $H \otimes h$ is a self-dual system, like $G \otimes g$, so we know that the bases and the values of the $\{h_i\}$ at the symmetry points will be multiples of the same vectors. Thus the electronic basis vectors at the D_{3d} and D_{2h} points are obtained by normalizing the rows of Tables G.5 and G.11 respectively. In addition we must include in the list the $\{h_i\}$ corresponding to a D_{5d} distortion, which is also allowed. This was not listed under $G \otimes h$ because there it was a degeneracy. This D_{5d} set of $\{h_i\}$ is simple to write down because the $\{h_i\}$ lie on the $\alpha = 0$ spherical subspace in $\{h_i\}$, at the vertices of an inscribed icosahedron, just as the D_{3d} lie on the vertices of the related inscribed dodecahedron. These $\{h_i\}$ are also tabulated in Appendix G, Table G.14, and the bases are obtained from them by normalizing the rows of the table.

As a result of putting all these bases into the Öpik and Pryce equations derived from this Jahn-Teller interaction, (5.3), we get the following sets of stationary points and energies:

Symmetry	Type	Energy	
D_{5d}	minima	$-54(k_{h,4}^H)^2$	
D_{2h}	Type I saddles	$-40(k_{h,4}^H)^2$	(5.4)
D_{2h}	Type II saddles	$-(45/2)(k_{h,4}^H)^2$	
D_{3d}	saddles	$-(50/3)(k_{h,4}^H)^2,$	

where the Type I and Type II D_{2h} points are the two sets distinguished by Ceulemans and Fowler (1990) (see Section 5.8).

The geometry in coordinate space associated with (5.4) is sketched schematically in Figure 5.1. The D_{5d} minima are shown at the vertices of a regular icosahedron inscribed in the $\alpha = 0$ spherical subspace, and lines of degeneracy

Figure 5.1. A closed path along edges of an icosahedron through D_{5d} minima, which are shown as spots. One line of degeneracies is shown as * at the center of the triangular path, and saddle points near the path are marked **S**.

pass through the D_{3d} points at the centers of the triangular faces. The lowest saddles, D_{2h}, type I, are not actually in the three-space in which the sphere is embedded, but there is one near the center of each line joining a pair of minima. This geometry is the inverse of that shown in Figure 4.2 for the $G \otimes h$ system.

To find the ground states, we need to know that the Berry phase changes over tunneling paths connecting the minima, and each tunneling path will pass close to the D_{2h} saddle point that lies near to the direct line. Tracking the Berry phase numerically along edges of the icosahedron shows that the phase changes by π around every triangular path such as the one shown in Figure 5.1. The method of numerical phase tracking is described by Cullerne and O'Brien (1994). Such direct testing of the existence of phase changes is necessary; the appearance of degeneracies within the paths does not guarantee a phase change, as Section 5.3 will illustrate.

This phase change of π around each triangular path will be correctly represented by a sign change between every pair of adjacent minima. There are twelve vertices on an icosahedron, but vertices related by inversion are identical points in phase space; thus there are six distinct D_{5d} minima, and each minimum is a nearest neighbor of every other. (Note that next nearest neighbors on the icosahedron are mapped by inversion to nearest neighbors). The paths between vertices related by inversion contain an odd number of neighbor-neighbor links, so inversion will change the sign of the electronic part of the wave function, $u(\mathbf{r}, Q)$. Thus to make the total wave function $\Psi = u(\mathbf{r}, Q)\psi(q)$ single-valued, the vibrational wave function $\psi(Q)$ must also change sign under inversion, and it also is assumed to change sign between every pair of minima. Thus the matrix of overlaps between states localized at the six minima can be written

$$S_{h,4} = \begin{bmatrix} 0 & -S & -S & -S & -S & -S \\ -S & 0 & -S & -S & -S & -S \\ -S & -S & 0 & -S & -S & -S \\ -S & -S & -S & 0 & -S & -S \\ -S & -S & -S & -S & 0 & -S \\ -S & -S & -S & -S & -S & 0 \end{bmatrix}, \tag{5.5}$$

where the negative signs follow from the change of sign of the wave function between every pair of minima (see Appendix B). This matrix has a single root at $-5S$, and a fivefold degenerate root at S. Because the effect of increasing the overlap is to decrease the tunneling energy, the latter set is the ground vibronic state, of H symmetry type. Its eigenvectors are the five columns of the array whose rows are the $\{h_i\}$ at the minima (Table G.14).

In this strong coupling ground state, the Ham reduction factor for a G-type operator is identically zero: $K(G) = 0$. This can be seen with reference to the matrix $M^H(g)$ in the H ⊗ g section (Section 5.5). One D_{5d} base is $\mathbf{u} = (1, 0, 0, 0, 0)$, and $M^H(g).\mathbf{u} = 0$, so $\langle u|M^H(g)|u\rangle = 0$. Because of the duplication of H ⊗ h matrices, the other Ham factors are given by a matrix equation, as described by Cullerne et al. (1995) (see Section 2.6). This equation can be written in terms of the matrices $M_2^H(h)$ and $M_4^H(h)$ with $k_2^H = k_4^H = 1$,

$$\begin{pmatrix} \langle M_2^H(h)\rangle_{\text{vib}} \\ \langle M_4^H(h)\rangle_{\text{vib}} \end{pmatrix} = \begin{pmatrix} \dfrac{2}{7} & \dfrac{3\sqrt{3}}{49} \\ \dfrac{12\sqrt{3}}{5} & \dfrac{18}{35} \end{pmatrix} \times \begin{pmatrix} M_2^H(h) \\ M_4^H(h) \end{pmatrix}. \tag{5.6}$$

This tells us that if an H-type operator has an interaction of the form $M_n^H(h)$ in the original electronic states, then its matrix in the vibronic ground states at strong coupling, $\langle M_n^H(h)\rangle_{\text{vib}}$, is a linear combination of the two matrices $M_2^H(h)$ and $M_4^H(h)$. This equation differs from the one given by Cullerne et al. because their matrices $M_n^H(h)$ have a different normalization. The diagonal factors are independent of normalization, as are simple Ham factors, being defined as a ratio, but the off-diagonal ones depend on the ratio of the choice of normalization of the two matrices, $M_2^H(h)$ and $M_4^H(h)$.

5.3 H ⊗ h IN GENERAL

This coupling gives rise to a lowest APES that either has minima at D_{5d} distortions, as for H ⊗ h_4 (Section 5.2), or at D_{3d} distortions, or with a continuous minimum over a five-dimensional hypersphere (Pooler's case [3], in [5.2]). We shall approach the most general form of the H ⊗ h coupling by using the two

Ceulemans and Fowler matrices. These matrices can be written as

$$M_a^H(h) = k_a^H \times \qquad (5.7)$$

$$\begin{bmatrix} 0 & \sqrt{3}h_2 & -\sqrt{3}h_3 & -\sqrt{3}h_4 & \sqrt{3}h_5 \\ \sqrt{3}h_2 & \sqrt{3}h_1 + h_4 & h_3 + h_5 & h_2 - h_4 & h_3 \\ -\sqrt{3}h_3 & h_3 + h_5 & -\sqrt{3}h_1 + h_2 & h_5 & h_2 + h_4 \\ -\sqrt{3}h_4 & h_2 - h_4 & h_5 & -\sqrt{3}h_1 - h_2 & h_3 - h_5 \\ \sqrt{3}h_5 & h_3 & h_2 + h_4 & h_3 - h_5 & \sqrt{3}h_1 - h_4 \end{bmatrix}$$

and

$$M_b^H(h) = k_b^H \times \qquad (5.8)$$

$$\begin{bmatrix} \frac{4h_1}{\sqrt{3}} & -\frac{h_2}{\sqrt{3}} & -\frac{h_3}{\sqrt{3}} & -\frac{h_4}{\sqrt{3}} & -\frac{h_5}{\sqrt{3}} \\ -\frac{h_2}{\sqrt{3}} & -\frac{h_1}{\sqrt{3}} + h_4 & -h_3 + h_5 & h_2 + h_4 & h_3 \\ -\frac{h_3}{\sqrt{3}} & -h_3 + h_5 & -\frac{h_1}{\sqrt{3}} - h_2 & -h_5 & h_2 - h_4 \\ -\frac{h_4}{\sqrt{3}} & h_2 + h_4 & -h_5 & -\frac{h_1}{\sqrt{3}} + h_2 & -h_3 - h_5 \\ -\frac{h_5}{\sqrt{3}} & h_3 & h_2 - h_4 & -h_3 - h_5 & -\frac{h_1}{\sqrt{3}} - h_4 \end{bmatrix}.$$

$M_a^H(h)$ on its own gives the D_{3d} minima at an energy $-(8/3)(k_a^H)^2$, as $M_b^H(h)$ likewise gives the D_{5d} minima at an energy $-(8/3)(k_b^H)^2$.

The most general $H \otimes h$ interaction matrix can now be written as

$$M_\beta^H(h) = k^H[\cos\beta \times M_b^H(h)/k_b^H + \sin\beta \times M_a^H(h)/k_a^H], \qquad (5.9)$$

and allowing β to cover a range of 2π covers all the possibilities for $H \otimes h$.

The range of possibilities can be explored with reference to Figure 5.2, where the eigenvalues of $M_\beta^H(h)$ at the points corresponding to normalized distortions of D_{5d} and D_{3d} symmetry are plotted as functions of β. These two distortions are chosen for investigation because the minima of the lowest APES will be at one or other of them, as will be the degeneracies. The eigenvalues with the largest absolute value of the energy are selected by the Jahn-Teller interaction to be on the lowest APES[1]. The figure shows how the D_{5d} distortions are selected over half the range of β, with D_{3d} distortions selected over the other half. The epikernel principle tells us that other distortions will not be selected, except at the special values $\beta = n\pi/4$, where there is a degeneracy between positive and negative eigenvalues as well as between D_{5d} and D_{3d} eigenvalues, leading to a degeneracy in the energy of the resulting distortions[2].

---

[1] This is because if $\{\lambda_i\}$ is a set of eigenvalues corresponding to a normalized distortion, then the energy of the lowest APES at that point is found by minimizing the energy $\lambda_i Q + Q^2/2$ with respect to the choice of i as well as with respect to the magnitude of the distortion Q.

[2] "Positive" distortions correspond to increases in the magnitudes of the normal coordinates; "negative" distortions correspond to decreases.

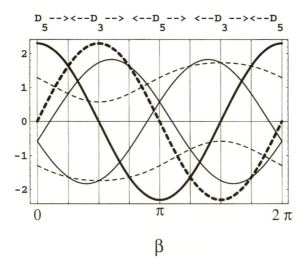

Figure 5.2. The eigenvalues of $M_\beta^H(h)$ at D_{5d} and D_{3d} points plotted against β. Solid lines are D_{5d} eigenvalues; dashed lines are D_{3d} eigenvalues. In both symmetries the heavy lines indicate non-degenerate states and the light lines indicate doubly degenerate states. The ranges of appearance of the two types of minima on the lowest APES are indicated.

The selected eigenvalues lead to a non-degenerate APES at the energy minima (if they did not, the Jahn-Teller theorem tells us that a further distortion would occur to remove this degeneracy) and the figure can be used to find some of the degeneracies on the lowest APES. For instance at $\beta = 0$, where the minima of energy are at D_{5d} points, corresponding to a positive eigenvalue, the highest state at D_{3d} is doubly degenerate, corresponding to a degeneracy in the lowest APES at that point. On the other hand when $\pi/(4\sqrt{3}) < \beta < \pi/4$, the highest eigenstate at D_{3d} is non-degenerate. Numerical phase tracking confirms that when there is a degeneracy on the lowest APES on the $\alpha = 0$ surface and the D_{5d} minima are selected, then there is a phase change of π over a path surrounding it, while the absence of a degeneracy goes with the absence of a phase change. In a similar way the figure shows that in part of the range where the D_{3d} minima are selected, there is a degeneracy on the lowest APES at the D_{5d} point; however, numerical phase tracking finds no phase change around this degeneracy.

The transition points where the D_{5d} and D_{3d} distortions have equal energy correspond to Pooler's case (3) in (5.2), to be discussed in Section 5.4. Putting $\beta = \pi/4$ or $\beta = 5\pi/4$ returns the matrix $M_2^H(h)$ or its negative. Putting $\beta = 3\pi/4$ or $\beta = 7\pi/4$ returns an interaction matrix that is changed to $M_2^H(h)$ by the same permutation of the $\{h_i\}$ as is used in Chapter 3 (equation [3.15]).

Thus the interaction given by these values of β will have an exactly similar SO(3) invariance to that of $H \otimes h_2$, but corresponding to a different SO(3) subspace. The case of the interaction matrix $M_4^H(h)$ discussed in Section 5.2 is recovered by putting $\tan \beta = -5/9$, that is $\beta = -0.16\pi$.

To find the nature of the ground states at strong coupling, we need as usual to look at the phase changes of the electronic wave function over closed tunneling paths on the lowest APES (paths which minimize the tunneling integral (B.11)); this has been done numerically. In ranges approximately $-\pi/5 < \beta < \pi/5$ and $4\pi/5 < \beta < 6\pi/5$, the discussion in Section 5.2 holds: there is a phase change of π between the D_{5d} minima which produces a quintet ground state. If the minima are at D_{5d}, but there is no phase change in the electronic wave function around any closed tunneling path, then the electronic wave function $u(\mathbf{r}, Q)$ does not change sign between adjacent minima, and the vibrational part $\psi(Q)$ must also keep the same sign between adjacent minima to preserve the single-valuedness of the vibronic function Ψ. The overlap matrix between states localized at the six minima can thus be taken as

$$
\begin{bmatrix}
0 & S & S & S & S & S \\
S & 0 & S & S & S & S \\
S & S & 0 & S & S & S \\
S & S & S & 0 & S & S \\
S & S & S & S & 0 & S \\
S & S & S & S & S & 0
\end{bmatrix} . \tag{5.10}
$$

This is just the negative of the matrix $S_{h,4}$, equation (5.5), and the eigenvalues will be the negative of those of $S_{h,4}$. The eigenstate of lowest energy, the one that maximizes the tunneling energy, is now the totally symmetric, non-degenerate A state.

When it comes to the D_{3d} minima, the situation is different. In the whole of the regions $(\pi/4 < \beta < 3\pi/4)$ and $(5\pi/4 < \beta < 7\pi/4)$, there is never any phase change around a closed tunneling path. This absence of phase change leads to a singlet A ground state again. The paths in this case are around the faces of a dodecahedron, as shown in Figure 4.2, and in some part of the range there is a degeneracy at the center of each face, which makes it surprising that no phase change is found.

This absence of a phase change makes it instructive to have another look at the working of Berry's theorem in this case (see Section 1.4 and Berry, 1984). Berry's analysis of the phase change shows explicitly that for a degeneracy with a Hamiltonian of the type shown in (5.11), there is a phase change of π as the electronic state is taken around any closed path encircling the degeneracy. In an analogous situation, the $E \otimes e$ Jahn-Teller interaction in cubic symmetry, Ham (1987) pointed out the connection of the degeneracy on the APES with the appearance of a degeneracy in the ground state of the Jahn-Teller system.

It is natural to ask why a phase change of π is not found in a path through D_{3d} minima encircling a degeneracy.

To see what is going on, we can have a look at the interaction matrix $M_a^H(h)$, (5.7), corresponding to $\beta = \pi/2$. Bearing in mind that the D_{5d} and D_{3d} points can all be plotted on the $\alpha = 0$ spherical surface, we concentrate on the D_{5d} point at the $\theta = 0$ point on that sphere and the pentagon of D_{3d} points surrounding it. At this particular D_{5d} point, the only nonzero coordinate is h_1, and nearby h_2, h_3, h_4, and h_5 are small. The lowest roots of $M_a^H(h)$ at this $\theta = 0$ point correspond to the second and fifth electronic bases, so the linear interaction between these bases near that point can be extracted from $M_a^H(h)$ in the form

$$M^{(1)} = \begin{bmatrix} -h_4 & -h_3 \\ -h_3 & h_4 \end{bmatrix}, \tag{5.11}$$

where a diagonal term $-\sqrt{3}h_1$ has been subtracted. This matrix is in the standard form for an $E \otimes \epsilon$ type of degeneracy (see Appendix C) where the ϵ modes are represented by the pair of coordinates $\{h_3, h_4\}$. It induces a phase change of π in the eigenstates on going around a circle that encloses the origin at $(0, 0)$. The other pair of modes, $\{h_2, h_5\}$, only appear in this interaction to second order. These second-order terms can be found by second-order perturbation mixing in the other roots of $M_a^H(h)$, and this part of the interaction, after subtracting $-2(h_2^2 + h_5^2)$ from the diagonal, can be written as

$$M^{(2)} = \begin{bmatrix} -\frac{3}{2}(h_2^2 - h_5^2) & -3h_2h_5 \\ -3h_2h_5 & \frac{3}{2}(h_2^2 - h_5^2) \end{bmatrix}. \tag{5.12}$$

$M^{(2)}$ is of the same form as a second order $E \otimes \epsilon^2$ Jahn-Teller interaction with the e modes now represented by the $\{h_2, h_5\}$ pair. Such an interaction does not produce a change of phase around any path in the $\{h_2, h_5\}$ plane. The difference between $M^{(1)}$ and $M^{(2)}$ can be seen by putting $h_4 = h_0 \cos \phi$ and $h_3 = h_0 \sin \phi$, so that one circuit of the degeneracy is produced by taking ϕ from 0 to 2π and

$$M^{(1)} = h_0 \begin{bmatrix} -\cos \phi & -\sin \phi \\ -\sin \phi & +\cos \phi \end{bmatrix}. \tag{5.13}$$

The eigenstates of $M^{(1)}$ are $[\cos(\phi/2), \sin(\phi/2)]$, and $[-\sin(\phi/2), \cos(\phi/2)]$, and both change sign under one circuit of the degeneracy at the origin in the $\{h_2, h_5\}$ plane. On the other hand, putting $h_2 = h_0 \cos \chi$, $h_5 = h_0 \sin \chi$ gives

$$M^{(2)} = h_0^2 \frac{3}{2} \begin{bmatrix} -\cos 2\chi & -\sin 2\chi \\ -\sin 2\chi & +\cos 2\chi \end{bmatrix}, \tag{5.14}$$

and the presence of the double angle ensures that the eigenstates are functions of $\cos \chi$, $\sin \chi$ so that there is no phase change upon going around the degeneracy.

Now the path around the edges of the pentagon can be transformed continuously and without crossing any degeneracy into a path that goes tightly around the degeneracy at $\theta = 0$ on the spherical surface. At this point the spherical surface is tangential to the $\{h_2, h_5\}$ plane and orthogonal to the $\{h_3, h_4\}$ plane. As we have seen, a path in the $\{h_2, h_5\}$ plane induces no phase change, and consequently it is reasonable that no phase change is induced by traversing the pentagonal path.

5.4 $H \otimes h_2$

5.4.1 Bases and the Hamiltonian

In this section we analyze the special case of $H \otimes h$ when the interaction matrix is the one derived from a set of $L = 2$ states, $M_2^H(h)$. The normal mode coordinates $\{h_i, \ i = 1, 2, \ldots 5\}$ are identical with the $\{Q_i\}$ used in Chapter 3, and we shall be using the angular parametrization given in equation (3.5). (The notation differs but the identification is simple: $\{h_i\} \equiv \{Q_i\}$). The complete Hamiltonian for this linear $H \otimes h_2$ system can be written as

$$\mathcal{H} = -\frac{1}{2} \sum_{i=1}^{5} \left(\frac{\partial}{\partial h_i} \right)^2 + \frac{1}{2} \sum_{i=1}^{5} (h_i)^2 + M_2^H(h). \tag{5.15}$$

We again are scaling the energies in units of $\hbar\omega$, and the ω is to be understood in terms of an effective frequency (discussed further in Chapter 6 in terms of the cluster model approximation). Appendix E (E.13) provides an expression for the interaction matrix $M_2^H(h_2)$.

$$M_2^H(h) = k_{h,2}^H \begin{bmatrix} 2h_1 & h_2 & -2h_3 & -2h_4 & h_5 \\ h_2 & h_1 + \sqrt{3}h_4 & \sqrt{3}h_5 & \sqrt{3}h_2 & \sqrt{3}h_3 \\ -2h_3 & \sqrt{3}h_5 & -2h_1 & 0 & \sqrt{3}h_2 \\ -2h_4 & \sqrt{3}h_2 & 0 & -2h_1 & -\sqrt{3}h_5 \\ h_5 & \sqrt{3}h_3 & \sqrt{3}h_2 & -\sqrt{3}h_5 & h_1 - \sqrt{3}h_4 \end{bmatrix}$$

$$\tag{5.16}$$

The constant $k_{h,2}^H$ scales the strength of the interaction between the phonon modes and the electronic quintet.

5.4.2 Rotational Symmetry of $H \otimes h_2$

The present system involves the same phonon modes first introduced in connection with the $T \otimes h$ systems in Chapter 3. As then, the reparametrization of the coordinates makes it easy to see the rotational invariance of the Hamiltonian. Given the same phonon modes as in $T \otimes h$, the oscillator Hamiltonian is the same and thus has the same appearance in terms of the $\{Q, \alpha, \gamma, \theta, \phi\}$ as shown in (3.19) and (3.20).

$$\mathcal{H}_{osc} = -\frac{1}{2}\left[Q^{-4}\frac{\partial}{\partial Q}\left(Q^4\frac{\partial}{\partial Q}\right) + [Q^2\sin(3\alpha)]^{-1}\frac{\partial}{\partial \alpha}\left(\sin(3\alpha)\frac{\partial}{\partial \alpha}\right)\right]$$

$$+\frac{1}{8Q^2}\left[\frac{\lambda_x^2}{\sin^2(\alpha - 2\pi/3)} + \frac{\lambda_y^2}{\sin^2(\alpha + 2\pi/3)} + \frac{\lambda_z^2}{\sin^2\alpha}\right] + \frac{1}{2}Q^2,$$

$$(5.17)$$

where $\{\lambda_x, \lambda_y, \lambda_z\}$ are the three components of an angular momentum operator λ within the phonon space. Bohr and Mottelson (1975) have dealt with an identical oscillator Hamiltonian in their analysis of quadrupole oscillations in nuclei. The present case is more complex due to the electron-phonon coupling contained within $M_2^H(h)$.

The SO(3) symmetry of the $H \otimes h_2$ Hamiltonian tells us that we can simplify the Hamiltonian equation through a judicious choice of unitary transformations. Our experience with $T \otimes h$ in Chapter 3 also should prompt us to look for rotations parametrized by each of the four coordinate angles $(\alpha, \theta, \phi, \gamma)$. To this end, appropriate unitary transformations can be developed which represent rotations in the quintet electronic space, with each rotation parametrized by one of the four angles.

These transformations and their angular dependences, explicitly given in Appendix F, are denoted as follows: $A_D(\alpha), B_D(\gamma), C_D(\theta),$ and $D_D(\phi)$. Using these unitary operators, $M_2^H(h)$ can be rewritten as follows:

$$2k_{h,2}^H Q\,T^{-1}\begin{bmatrix} 1 & 0 & 0 & 0 & \\ 0 & \cos(\alpha - \pi/3) & 0 & 0 & 0 \\ 0 & 0 & -\cos\alpha & 0 & 0 \\ 0 & 0 & 0 & -1 & 0 \\ 0 & 0 & 0 & 0 & \cos(\alpha + \pi/3) \end{bmatrix} T \quad (5.18)$$

The generalized rotation operator $T_D = A_D(\alpha)B_D(\gamma)C_D(\theta)D_D(\phi)$ incorporates two actions: (1) a rotation through the Euler angles (γ, θ, ϕ) and (2) a rotation in α.

As (5.18) shows, the eigenvalues of the interaction Hamiltonian, $M_2^H(h)$, are surprisingly easy to express,

$$\{+2Q, -2Q\cos(\alpha - \pi/3), -2Q\cos\alpha, -2Q, Q\cos(\alpha + \pi/3)\} \quad (5.19)$$

as ordered from top left in the matrix, each multiplied by $k_{h,2}^H$. In the adiabatic approximation, solutions to (5.15) involve one of five potential energy surfaces given by the eigenvalues of $M_2^H(h)$ with the addition of the restoring term $Q^2/2$. The lowest of these

$$U_4 = \frac{1}{2}Q^2 - 2k_{h,2}^H Q \tag{5.20}$$

is a function of Q only, with a minimum value of $-2k_{h,2}^H$ at $Q = Q_0 \equiv 2k_{h,2}^H$. The lowest APES thus forms a four-dimensional hypersphere of radius Q_0 in the five-dimensional space of the vibrational coordinates $\{Q, \alpha, \gamma, \theta, \phi\}$. The appearance of this surface is proof of the SO(5) symmetry of the ground state.

The APES corresponding to the $-2k_{h,2}^H Q \cos \alpha$ eigenvalue,

$$U_3 = \frac{1}{2}Q^2 - 2k_{h,2}^H Q \cos \alpha, \tag{5.21}$$

is at its minimum value for $Q = Q_0$ and $\alpha = 0$. The minima of U_2 and U_5 thus are both $-2(k_{h,2}^H)^2$, and the two APESs will be degenerate over a three-dimensional subspace of the hypersphere defined by $\alpha = 0$. The remaining APESs,

$$U_1 = \frac{1}{2}Q^2 + 2k_{h,2}^H Q \tag{5.22}$$

$$U_2 = \frac{1}{2}Q^2 + 2k_{h,2}^H Q \cos(\alpha - \pi/3) \tag{5.23}$$

$$U_5 = \frac{1}{2}Q^2 - 2k_{h,2}^H Q \cos(\alpha + \pi/3) \tag{5.24}$$

do not coincide in energy and are degenerate with the lowest APES only at $Q = 0$.

5.4.3 The Ground States at Strong Coupling

The column vectors of the $\mathcal{T}_D$ matrix define the eigenvectors for states associated with each of the APESs. For U_4, the lowest of the five APESs, the corresponding eigenvector is

$$|u_4\rangle = \sin(\alpha/2)\mathbf{a}_1(\gamma, \theta, \phi) + \cos(\alpha/2)\mathbf{a}_2(\gamma, \theta, \phi), \tag{5.25}$$

where

$$\mathbf{a}_1(\gamma, \theta, \phi) = \begin{bmatrix} \frac{1}{2}(3\cos^2\theta - 1) \\ \frac{\sqrt{3}}{2}\sin 2\theta \cos\phi \\ \frac{\sqrt{3}}{2}\sin^2\theta \sin 2\phi \\ \frac{\sqrt{3}}{2}\sin^2\theta \cos 2\phi \\ \frac{\sqrt{3}}{2}\sin 2\theta \sin\phi \end{bmatrix} \tag{5.26}$$

and

$$
\mathbf{a}_2(\gamma,\theta,\phi) = \begin{bmatrix}
\frac{\sqrt{3}}{2}\sin^2\theta\cos 2\gamma \\
-\frac{1}{2}\sin 2\theta\cos\phi\cos 2\gamma + \sin\theta\sin\phi\sin 2\gamma \\
\frac{1}{2}(1+\cos^2\theta)\sin 2\phi\cos 2\gamma + \cos\theta\cos 2\phi\sin 2\gamma \\
\frac{1}{2}(1+\cos^2\theta)\cos 2\phi\cos 2\gamma - \cos\theta\sin 2\phi\sin 2\gamma \\
-\frac{1}{2}\sin 2\theta\sin\phi\cos 2\gamma - \sin\theta\cos\phi\sin 2\gamma
\end{bmatrix}. \quad (5.27)
$$

We note that in terms of the vectors $\mathbf{a}_1$ and $\mathbf{a}_2$, the parametrization of the $\{h_i\}$ can be written

$$
\mathbf{h} = Q\left(\cos\alpha\,\mathbf{a}_1(\gamma,\theta,\phi) + \sin\alpha\,\mathbf{a}_2(\gamma,\theta,\phi)\right). \quad (5.28)
$$

The eigenvector associated with the highest APES, U_1, similarly takes the form

$$
|u_1\rangle = \cos(\alpha/2)\mathbf{a}_1(\gamma,\theta,\phi) - \sin(\alpha/2)\mathbf{a}_2(\gamma,\theta,\phi). \quad (5.29)
$$

The three intermediate roots have eigenvectors that do not depend on α. They are as follows:

$$
\begin{aligned}
|u_2\rangle &= \mathbf{a}_3(\gamma,\theta,\phi) \quad (5.30) \\
|u_3\rangle &= \mathbf{a}_2\left(\gamma + \frac{\pi}{4},\theta,\phi\right) \\
|u_5\rangle &= \mathbf{a}_3\left(\gamma + \frac{\pi}{2},\theta,\phi\right),
\end{aligned}
$$

where

$$
\mathbf{a}_3(\gamma,\theta,\phi) = \begin{bmatrix}
-\frac{\sqrt{3}}{2}\sin 2\theta\cos\gamma \\
\cos 2\theta\cos\phi\cos\gamma - \cos\theta\sin\phi\sin\gamma \\
\frac{1}{2}\sin 2\theta\sin 2\phi\cos\gamma + \sin\theta\cos 2\phi\sin\gamma \\
\frac{1}{2}\sin 2\theta\cos 2\phi\cos\gamma - \sin\theta\sin 2\phi\sin\gamma \\
\cos 2\theta\sin\phi\cos\gamma + \cos\theta\cos\phi\sin\gamma
\end{bmatrix}. \quad (5.31)
$$

The eigenvectors for the lowest- and highest-energy APESs, $|u_4\rangle$ and $|u_1\rangle$, are the only ones that contain $\cos(\alpha/2)$ and $\sin(\alpha/2)$ terms. This half-angle dependence ensures that distortions related by an inversion in $\{h_i\}$ space appear on the lowest APES with different electronic eigenstates.

The overlap between the U_4 and U_3 APESs defined by setting $\alpha = 0$ creates an ambiguity in the choice of eigenvectors for this region, but $|u_4\rangle$ and $|u_3\rangle$ are properly orthogonal, and they change continuously through $\alpha = 0$. Thus if we ignore any breakdown in the adiabatic approximation where the surfaces meet, the vibronic Hamiltonian can be solved separately on each APES.

In obtaining approximate strong-coupling ground state wave functions, we assume the adiabatic approximation and start with a Born-Oppenheimer product of states,

$$
\Psi = \psi(h,\alpha,\gamma,\theta,\phi)\,|u_4\rangle, \quad (5.32)
$$

as a solution to the strong-coupling Hamiltonian

$$(\mathcal{H}_0 + U_4) \, \Psi = E \Psi, \tag{5.33}$$

where $\mathcal{H}_0$ is given in (5.17). The vibrational part of the wave function, ψ, is obtained by substituting (5.32) into (5.33) to get an equation for ψ. This equation can be split into two independent equations by a separation of variables using $\psi = f(Q) \, \Phi(\alpha, \, L, \, M)$, where L and M label the λ angular momentum state and its azimuthal component. The equation in Q can be written as

$$\frac{d^2 F}{d Q^2} + \frac{(\Lambda - 9/4)}{Q^2} F - (Q - 2k_{h,2}^H)^2 F + (2 \, E + 4k_{h,2}^H) \, F = 0, \tag{5.34}$$

using $F = f/Q^2$. The equation for Φ is

$$\frac{1}{\sin(3\alpha)} \frac{d}{d\alpha} \left(\sin(3\alpha) \frac{d}{d\alpha} \right) \Phi$$

$$-\frac{1}{4} \left[\frac{\lambda_x^2}{\sin^2(\alpha - 2\pi/3)} + \frac{\lambda_y^2}{\sin^2(\alpha + 2\pi/3)} + \frac{\lambda_z^2}{\sin^2 \alpha} \right] \Phi = \Lambda \, \Phi. \tag{5.35}$$

Equation (5.34) describes a one-dimensional oscillation normal to the surface of the hypersphere. For low-energy motions we can put $Q = Q_0 \equiv 2k_{h,2}^H$, which gives

$$E \approx -2(k_{h,2}^H)^2 + \frac{1}{2} - \frac{(\Lambda - 9/4)}{4 \, (k_{h,2}^H)^2}, \tag{5.36}$$

including the zero-point energy. The allowed values for Λ are determined by noting that acceptable solutions to (5.35) require that $\Lambda = -\ell(\ell + 3)$, with ℓ a positive integer or zero. This is a general result, given a hyperspherically symmetric potential in a five-dimensional space and the requirement that Φ be single-valued in coordinate space. The integer ℓ is the five-dimensional analogue of the angular momentum quantum number in three dimensions and it indexes the symmetric irreps of SO(5): $[\ell, 0]$. With $L = 0$, $\ell = 0, 3, 6, \ldots$, and if $L = 2$ then $\ell = 1, 2, 4, 5, 7, \ldots$; there are no $L = 1$ states, a result that follows from a symmetry analysis of the five-dimensional harmonic oscillator. (Explicit expressions for Φ for the lowest few values of L are derived in Chancey and O'Brien, [1989]). The state of lowest energy is the singlet with $\ell = 0$, $L = 0$.

The requirement that the complete vibronic wave function Ψ be single-valued is preserved in this case because $|u_4\rangle$ is invariant under the symmetry operations of the quadrupole oscillator that rotate the oscillator through one complete revolution (see Eisenberg and Greiner (1970), Section 4.1), and so is also single-valued in coordinate space. This invariance is at first surprising, given the apparent analogy of the $\{|u_4\rangle, |u_1\rangle\}$ subsystem with the simpler $E \otimes e$ system of cubic symmetry: Both the electronic ground states possess a similar half-angle dependence. The $E \otimes e$ electronic ground state, however, is transformed to the negative of itself, and so is not single valued in its coordinate space. This difference leaves open the door for the appearance of the SO(5) [0, 0] ground state in $H \otimes h_2$.

5.5 $H \otimes g$

The Jahn-Teller interaction matrix for this system is given in Appendix E, (E.15), and it is

$$
M^H(g) = k_g^H \times \begin{bmatrix}
0 & 2\sqrt{3}g_3 & 2\sqrt{3}g_2 & -2\sqrt{3}g_1 & 2\sqrt{3}g_4 \\
2\sqrt{3}g_3 & -4g_1 & g_2 - g_4 & g_1 - g_3 & 4g_2 \\
2\sqrt{3}g_2 & g_2 - g_4 & -4g_3 & -4g_4 & -g_1 - g_3 \\
-2\sqrt{3}g_1 & g_1 - g_3 & -4g_4 & 4g_3 & g_2 + g_4 \\
2\sqrt{3}g_4 & 4g_2 & -g_1 - g_3 & g_2 + g_4 & 4g_1
\end{bmatrix}.
$$
(5.37)

Some stationary points can be found using the method of Öpik and Pryce, and others related by symmetry can be deduced. The lowest energy is at a set of ten D_{3d} minima. At these points the set of values of the vibrational coordinates are a multiple of the electronic eigenvectors at the D_{3d} minima of $G \otimes h$, while, conversely, the electronic eigenvectors for $H \otimes g$ are a multiple of the set of vibrational coordinates at the minima of $G \otimes h$. This can be described as a duality between the two systems in which the electronic part and the vibrational part of the interaction are interchanged. Values are given in Appendix G (G.13).

In order to characterize the ground states in more detail, these ten minima have to be plotted in the four-dimensional $\{g_i\}$ space. In that space each minimum has six others as equidistant nearest neighbors and three at a greater distance. The topology is such that if the minima are plotted on the vertices of a dodecahedron, then these nearest neighbors are those that are *not* nearest neighbors on the dodecahedron; nearest neighbors on the dodecahedron are next-nearest neighbours in $H \otimes g$. Vertices of the dodecahedron related by inversion correspond to the same minimum as in Figure 4.2. With six nearest neighbors, the network of lines in the $\{g_i\}$ space along which the phase needs to be tracked is more complicated than was the case for $G \otimes g$, but the result is similar: phase changes on closed loops are correctly represented if there is assumed to be a sign change of the base between every pair of neighbors. The overlap matrix can thus be derived from the overlap matrix for the ground states of $G \otimes h$ (Section 4.4.1) by replacing every off-diagonal zero with $-S$ and every $-S$ with zero. The eigenvectors of this matrix are the same as for the $G \otimes h$ matrix, but the energies come out in a different order, with H lowest as might be expected, then G and then A.

The energies at all the special points, including the minima discussed above, can be found by substituting suitable bases into the appropriate Öpik and Pryce equations. These energies are as follows:

Symmetry	Type	Energy
D_{3d}	minima	$-(32/3)(k_g^H)^2$
T_d	Type I degeneracies	$-4(k_g^H)^2$
T_d	Type II degeneracies	$-9(k_g^H)^2$
D_{5d}	degeneracies	0

$$(5.38)$$

Note that with g vibrations alone, the sets of D_{2d} points collapse into sets of points of T_d symmetry.

The Ham factors are derived as described by Cullerne et al. (1995) as

$$K(G) = 4/9 \qquad (5.39)$$

together with the equation (in which $k_2^H = k_4^H = 0$)

$$\begin{pmatrix} \langle M_2^H(h) \rangle_{\text{vib}} \\ \langle M_4^H(h) \rangle_{\text{vib}} \end{pmatrix} = \begin{pmatrix} \frac{2}{7} & -\frac{2}{21\sqrt{3}} \\ -\frac{10}{7\sqrt{3}} & \frac{10}{63} \end{pmatrix} \times \begin{pmatrix} M_2^H(h) \\ M_4^H(h) \end{pmatrix}. \qquad (5.40)$$

As in Section 5.2, this equation differs from the one given by Cullerne et al. because their matrices $M_n^H(h)$ have a different normalization. The diagonal factors are independent of normalization, as are simple Ham factors, being defined as a ratio, but the off-diagonal ones depend on the ratio of the choice of normalization of the two matrices, $M_2^H(h)$ and $M_4^H(h)$.

5.6 H $\otimes$ $(g \oplus h_4)_{eq}$

The extra symmetry in this special case arises because g and h_4 together span an $L = 4$ representation of SO(3), while H spans $L = 2$. It would be possible to start afresh in this basis, but it is perhaps more instructive to build on what we already have.

Now that we have the two tables of energies, (5.38) and (5.4), for the systems $H \otimes g$ and $H \otimes h_4$, it is simple to put them together and find the parameters that give a uniform lowest energy surface. If we take

$$(k_g^H)^2 = \frac{7}{2}(k_{h,4}^H)^2, \qquad (5.41)$$

then the sum of the energies in the two tables is $-54(k_{h,4}^H)^2$ at all four special points. This then must be the ratio of coupling constants that gives Pooler's SO(3) invariance. We have not set up an explicit form for the Hamiltonian to express this invariance, unlike the case of $H \otimes h_2$, but it can be tracked numerically by the substitution of bases into the Öpik and Pryce equations.

The most general basis in the set of five electronic H states is a normalized five-component vector, and the Öpik and Pryce equations give the conditions that such components must satisfy at a stationary point of the APES. The surprising result is that all possible bases turn out to be solutions of these equations corresponding to the same energy. As a complete set of bases is a complete set of normalized five-component vectors, the search has been pursued over a hypersphere in five-dimensional space, and so the energy is invariant under SO(5). This result parallels that of the system $H \otimes h_2$, where the lowest APES of a Hamiltonian of SO(3) symmetry also is invariant under SO(5).

The lower symmetry of the Hamiltonian as a whole appears when we find the energies of the higher APESs at each point in configuration space above the minimum. These energies do vary for different choices of basis. However, if a basis is chosen belonging to any of the D_{5d}, D_{3d}, and D_{2h} points, the same pattern of APES energies is found, and we deduce that this is the set of energies to be found everywhere on the SO(3) invariant subspace. The APES energies at these points are

$$\{-54, -9, -9, 36, 36\} \times (k_{h,4}^H)^2. \tag{5.42}$$

This is again similar to the case $H \otimes h_2$ already discussed, where although the lowest energy has SO(5) invariance, higher energies are only invariant under SO(3). Here it is the condition holding α constant that holds the system on an SO(3) invariant continuum and keeps the higher energies constant.

5.7 $H \otimes (g \oplus h_4 \oplus h_2)_{eq}$

In the group chain $SO(5) \supset SO(3)$, the five-dimensional irrep $[1, 0]$ goes to $L = 2$ and the fourteen-dimensional irrep $[2, 0]$ goes to $L = 4 \oplus L = 2$, so this could be called a $[1, 0] \otimes [2, 0]$ Jahn-Teller system. It could be treated from scratch in this guise, but we shall build it up from its component parts. In putting together $H \otimes (g \oplus h_4)_{eq}$ and $H \otimes h_2$, the energy of the ground state is no help because it is already flat. It is necessary, therefore, to find the coupling constant that smooths all the higher energies. This can be done by a numerical search, and the magic ratio turns out to be

$$k_{h,2}^H = \frac{15}{\sqrt{14}} k_{h,4}^H. \tag{5.43}$$

With this ratio the energies are

$$\{-84, 21, 21, 21, 21\} \times (k_{h,4}^H)^2. \tag{5.44}$$

This system is the exact analogue of the system $G \otimes (g \oplus h)_{eq}$ discussed in Section 4.5.1. To get complete expressions for the pseudo-rotational energies, a parametrization of the fourteen components of the $[2, 0]$ irrep is needed, as is one for the five components of the $[1, 0]$, or $L = 2$ irrep. However, without going to this trouble, we can note that in five dimensions the angular parts of the solutions of the eigenvalue equation

$$-\frac{1}{2}\nabla^2\psi + V(q)\psi = E\psi \tag{5.45}$$

correspond to the SO(5) irreps $[\ell, 0]$, where ℓ is an integer. The ℓ-dependence of the energy is

$$V(\ell) = \frac{\ell(\ell + 3)}{Q_0^2}, \tag{5.46}$$

where Q_0 is the radius vector of the five-dimensional hyperspherical surface[3]. The electronic state will be constructed in the same way as it is for $G \otimes (g \oplus h)_{eq}$ (see equation [4.36]), and will be odd under inversion on the five-dimensional hypersphere. Consequently the vibronic state must also be odd under inversion, and the lowest odd state is the $[1,0]$ irrep with five components, which is of H type.

5.7.1 The Ham Reduction Factors

The symmetric square of the $[1, 0]$ irrep contains only $[2, 0]$ apart from the totally symmetric $[0, 0]$ irrep, so there is only one Ham factor, $K([2, 0])$, to be calculated. This Ham factor will apply equally to g operators and any h operator, which are all equally components of $[2, 0]$, so it can be calculated for any one of these fourteen operators. We shall proceed to do this for the operator h_1 in $M_2^H(h)$, which is conveniently diagonal.

Every electronic basis corresponds to a point on the Jahn-Teller minimum, so it can be taken quite generally as $(x_1, x_2, x_3, x_4, x_5)$, where the x_i can be thought of as the Cartesian coordinates of a point on a hypersphere of radius one. The expectation of one particular h_1 operator in this basis can be read off from the matrix $M_2^H(h)$ with $k_{h,2}^H$ set equal to one as

$$O_2(h_1) = 2x_1^2 + x_2^2 - 2x_3^2 - 2x_4^2 + x_5^2. \tag{5.47}$$

This is going to have to be taken as an operator within the vibronic states, averaged over the five-dimensional surface, and the set of vibronic states can be taken as $(x_1, x_2, x_3, x_4, x_5) \times \sqrt{5}$. The extra factor $\sqrt{5}$ here corresponds to the fact that the individual vibronic states are normalized to one, while

---

[3]This expression for the energy is less complete than that given for $H \otimes h_2$ (5.36) because here we do not have complete forms of the kinetic energy operator or of the electronic basis.

the normalization of the electronic basis is of the sum of the squares. The expectation of $O_2(h_1)$ within the first, or H_1, vibronic state is thus

$$\langle 1|O_2(h_1)|1\rangle = \langle 5(2x_1^4 + x_1^2 x_2^2 - 2x_1^2 x_3^2 - 2x_1^2 x_4^2 + x_1^2 x_5^2)\rangle \qquad (5.48)$$

where the symbols $\langle\ \rangle$ indicate an average over the five-dimensional hypersphere. As the x_i are just Cartesian coordinates, these averages are invariant under permutation of the indices. There are only two different averages to be calculated, which are

$$\langle x_1^4\rangle = \frac{3}{35}, \qquad \text{and} \qquad \langle x_1^2 x_2^2\rangle = \frac{1}{35}, \qquad (5.49)$$

so that

$$\langle 1|O_2(h_1)|1\rangle = \langle 10(x_1^4 - x_1^2 x_2^2)\rangle = \frac{4}{7}. \qquad (5.50)$$

When this is compared with the original matrix element, we see that

$$K([2, 0]) = \frac{2}{7}. \qquad (5.51)$$

Results such as (5.46) and (5.49) are best derived using the hyperspherical coordinates of Louck (1960). For five dimensions this parametrization can be taken as

$$x_1 = q \cos\theta_1, \qquad (5.52)$$
$$x_2 = q \cos\theta_2 \sin\theta_1,$$
$$x_3 = q \cos\theta_3 \sin\theta_2 \sin\theta_1,$$
$$x_4 = q \cos\theta_4 \sin\theta_3 \sin\theta_2 \sin\theta_1,$$
$$x_5 = q \sin\theta_4 \sin\theta_3 \sin\theta_2 \sin\theta_1.$$

This set of coordinates is orthogonal, and $x_1^2 + x_2^2 + x_3^2 + x_4^2 + x_5^2 = q^2$. The range of the variables is given by $0 \leq q < \infty$, $0 \leq \theta_1, \theta_2, \theta_3 < \pi$, and $0 \leq \theta_4 < 2\pi$. The element of area on the $q = 1$ hypersphere is

$$d\Omega_5 = \sin^3\theta_1 \sin^2\theta_2 \sin\theta_3 d\theta_1 d\theta_2 d\theta_3 d\theta_4, \qquad (5.53)$$

and the averages in (5.49) are given by

$$\langle x_1^4\rangle = \frac{\int x_1^4 d\Omega_5}{\int d\Omega_5} \quad \text{and} \quad \langle x_1^2 x_2^2\rangle = \frac{\int x_1^2 x_2^2 d\Omega_5}{\int d\Omega_5}. \qquad (5.54)$$

The expression for the contribution of the angular momentum to the energy, (5.46), is derived by Louck, and is to be found in his equation (75b). These variables are used by Ceulemans and Fowler (1990), who give a useful account of them in their appendix.

5.8 $H \otimes (g \oplus h)$

This most general case of linear coupling to a single g-mode and a single h-mode has been studied by Ceulemans and Fowler (1990), as mentioned in this chapter's introduction. They categorize their results in terms of three different Jahn-Teller stabilization energies, E_G^{JT}, the energy at the minima of $H \otimes g$ (5.38), together with E_{Ha}^{JT} and E_{Hb}^{JT}, which are the minimum energies corresponding to the two forms of the interaction matrices $M_a^H(h)$ (5.7) and $M_b^H(h)$ (5.8). With these energies they define an average Jahn-Teller stabilization energy,

$$E^0 = (4E_G^{JT} + 5E_{Ha}^{JT} + 5E_{Hb}^{JT})/14, \qquad (5.55)$$

and an energy parameter,

$$E^1 = 5(4E_G^{JT} + 5E_{Ha}^{JT} - 9E_{Hb}^{JT})/56, \qquad (5.56)$$

which is proportional to the difference between $9E_{Hb}^{JT}$ and $4E_G^{JT} + 5E_{Ha}^{JT}$.

Ceulemans and Fowler use an isostationary function method and show that the only stationary points correspond to D_{5d}, D_{3d}, and D_{2h} symmetry. The D_{2h} points are always saddle points, with D_{5d} minima if $E^1 > 0$ and D_{3d} minima if $E^1 < 0$. They identified two sets of D_{2h} saddle points, one a set of isolated points (called Type I in this chapter), and one a set situated on continuous curves in phase space (Type II).

The energy of the D_{3d} minima does not depend on k_b^H, and the D_{5d} energy is independent of k_a^H. They observe that $H \otimes (g \oplus h)$, in spite of its complexity, has the form of a two-mode problem, with separate trigonal and pentagonal coupling schemes in the a and b interaction matrices. The systems $H \otimes g$ (Section 5.5) and $H \otimes h$ (Section 5.3), $H \otimes h_2$ (Section 5.4) and $H \otimes h_4$ (Section 5.2), are special cases of this more general one.

The actual forms of the matrices listed by Ceulemans and Fowler differ from those in this book because they are prepared with an axis of quantization along a C_2 axis instead of along a C_5 axis as here.

6

Bridge to Experiment

INTRODUCTION

Our analysis of model systems in Chapters 3, 4, and 5 leads naturally to the question: How can the model results be integrated into an understanding of real icosahedral complexes? In attempting to answer this question, we shall first address the question of the validity of an obvious simplification in the model analyses, namely, the single-mode approximation. Afterwards, we shall deal directly with electron spin resonance (esr), with energy levels in anions of C_{60}, and with molecular spectra in $C_{60}^{\pm}$. We close with a brief discussion of superconductivity in the fullerides from the perspective of the $p^3 \otimes h$ model in Section 3.5.

6.1 MULTIMODE EFFECTS: CLUSTER MODELS

The theory developed so far has been in terms of a single set of vibrational modes belonging to each representation, while in many real contexts there will be several modes of each symmetry. All of them may well have Jahn-Teller coupling to any given electronic state. This situation is at its most extreme in solids, where the modes form a continuous distribution, but it is also important in large molecules such as C_{60}, where there will inevitably be several modes of each symmetry type. This proliferation of modes can be handled at weak coupling by perturbation methods: the calculations become complicated because excitations of all the modes involved have to be included in the list of excited states, but they can be done. At strong coupling there are a variety of cluster models that can be used. These are called cluster models because they were devised for use in solids, where it was assumed that a good first approximation might be to treat first the cluster of the ion and its immediate neighbors, and then to couple in the rest of the solid. These models involve making a canonical transformation to extract from the multimode Hamiltonian a part that has the same form as a single-mode one, and they will be discussed below. As usual the intermediate coupling strength regime offers the most difficult cases, and they have to be handled by judicious interpolation. The reasons for handling several modes in this way are twofold. The first reason is that it works: there is one choice of cluster that gives a very good representation of the ground states, both as regards their energies and their Ham factors, and there is another choice that gives a good representation of the overall band shape for optical absorption. The other reason is that the theory for single mode Jahn-Teller systems has already been worked out, and an approximation that starts with a cluster-model Hamiltonian already has known solutions. In fact, measurements of ground state properties or of broad band shapes will not distinguish between single-mode and multimode systems.

6.1.1 A Cluster Model for the Low Energy States

Most features of the cluster transformations can be brought out by considering the effects of only two sets of modes, so in search of simplicity we shall do just that.

6.1.1.1. A Cluster Hamiltonian for Two Modes

To simplify the manipulations, and in particular to avoid having to transform differential operators, we shall work in second quantization, the raising and lowering operators that tidy up any problem involving harmonic oscillators. For an oscillator whose Hamiltonian is

$$\mathcal{H}_{ph} = -\frac{\hbar^2}{2m}\frac{\partial^2}{\partial Q^2} + \frac{1}{2}m\omega^2 Q^2, \tag{6.1}$$

these are defined as

$$a = \sqrt{\frac{m\omega}{2\hbar}}Q + \sqrt{\frac{\hbar}{2m\omega}}\frac{\partial}{\partial Q} \tag{6.2}$$

$$a^\dagger = \sqrt{\frac{m\omega}{2\hbar}}Q - \sqrt{\frac{\hbar}{2m\omega}}\frac{\partial}{\partial Q},$$

so that

$$\mathcal{H}_{ph} = \hbar\omega\left(a^\dagger a + \frac{1}{2}\right) \tag{6.3}$$

and

$$Q = \sqrt{\frac{\hbar}{2m\omega}}(a + a^\dagger). \tag{6.4}$$

These operators have the property that if $|n\rangle$ is the nth harmonic oscillator state, of energy $(n + \frac{1}{2})\hbar\omega$, then

$$a^\dagger|n\rangle = \sqrt{n+1}|n+1\rangle \tag{6.5}$$
$$a|n\rangle = \sqrt{n}|n-1\rangle,$$

and they obey the commutation relations

$$[a, a^\dagger] = 1. \tag{6.6}$$

In terms of operators of this kind, and supposing there are two sets of modes of vibration corresponding to the same symmetry type ($i = 1, 2$), the vibrational Hamiltonian can be written (omitting the zero-point energy) as

$$\mathcal{H}_{ph} = \sum_{i=1}^{2} \sum_{j} \hbar\omega_i a_j^{(i)\dagger} a_j^{(i)}. \tag{6.7}$$

Here j labels the component of the vibrational mode within its symmetry representation and can be taken to be the same as the suffix already used in defining the components of the basis states. In this notation the Jahn-Teller interaction can be written

$$M = \sum_{i=1}^{2} \sum_{j} k_i \hbar\omega_i (a_j^{(i)} + a_j^{(i)\dagger}) U(j), \tag{6.8}$$

where the **U** matrices depend only on the symmetry (see Section 2.4). Now the aim is to find a linear combination of modes that, treated as a single mode on its own, will give as good an approximation as we can find to the two-mode Hamiltonian. This mode will then be the effective cluster mode. Accordingly we define

$$\begin{aligned} \alpha_j^{(\text{eff})} &= s_1 a_j^{(1)} + s_2 a_j^{(2)} \\ \alpha_j^{(\text{eff})\dagger} &= s_1 a_j^{(1)\dagger} + s_2 a_j^{(2)\dagger} \end{aligned} \tag{6.9}$$

with s_1 and s_2 real and $s_1^2 + s_2^2 = 1$. We shall need another pair of operators to transform the Hamiltonian, and for that we use the orthogonal combinations

$$\begin{aligned} \alpha_j^{(2)} &= s_2 a_j^{(1)} - s_1 a_j^{(2)} \\ \alpha_j^{(2)\dagger} &= s_2 a_j^{(1)\dagger} - s_1 a_j^{(2)\dagger}. \end{aligned} \tag{6.10}$$

The operator pair ($\alpha^{(\text{eff})}, \alpha^{(2)}$) will thus replace ($a^{(1)}, a^{(2)}$), and our goal is to choose $\alpha^{(\text{eff})}$ so as to maximize its effect. The conditions on the coefficients and orthogonality ensure that the commutation relations analogous to (6.6) are satisfied

$$[\alpha_j^{(i)}, \alpha_p^{(q)\dagger}] = \delta_{j,p}\delta_{i,q}. \tag{6.11}$$

The α's are well-behaved raising and lowering operators that generate harmonic oscillator states just as the original a's did. Making the change of variable produces the Hamiltonian in the form

$$\mathcal{H} = \mathcal{H}_{\text{eff}} + \mathcal{H}_{ph'} + \mathcal{H}_I, \tag{6.12}$$

with

$$\mathcal{H}_{\text{eff}} = \hbar \sum_j [(s_1^2 \omega_1 + s_2^2 \omega_2)\alpha_j^{(\text{eff})\dagger}\alpha_j^{(\text{eff})} \tag{6.13}$$

$$+ (k_1\omega_1 s_1 + k_2\omega_2 s_2)(\alpha_j^{(\text{eff})} + \alpha_j^{(\text{eff})\dagger})\mathbf{U}(j)],$$

$$\mathcal{H}_{ph'} = \hbar \sum_j [(s_2^2 \omega_1 + s_1^2 \omega_2)\alpha_j^{(2)\dagger}\alpha_j^{(2)}],$$

$$\mathcal{H}_I = \hbar \sum_j [(\omega_1 - \omega_2)s_1 s_2(\alpha_j^{(2)\dagger}\alpha_j^{(\text{eff})} + \alpha_j^{(\text{eff})\dagger}\alpha_j^{(2)})$$

$$+ (k_1\omega_1 s_2 - k_2\omega_2 s_1)(\alpha_j^{(2)} + \alpha_j^{(2)\dagger})\mathbf{U}(j)],$$

which has been divided so that $\mathcal{H}_{\text{eff}}$ is the cluster Hamiltonian, $\mathcal{H}_{ph'}$ is the Hamiltonian of another uncoupled harmonic oscillator, and $\mathcal{H}_I$ is the interaction between the eigenstates of $\mathcal{H}_{\text{eff}}$ and $\mathcal{H}_{ph'}$. Specifically, the parameters in the cluster Hamiltonian are

$$\omega_{\text{eff}} = (s_1^2 \omega_1 + s_2^2 \omega_2), \tag{6.14}$$

$$k_{\text{eff}}\omega_{\text{eff}} = (k_1\omega_1 s_1 + k_2\omega_2 s_2),$$

so that

$$\mathcal{H}_{\text{eff}} = \hbar\omega_{\text{eff}} \sum_j [\alpha_j^{(\text{eff})\dagger}\alpha_j^{(\text{eff})} + k_{\text{eff}}(\alpha_j^{(\text{eff})\dagger} + \alpha_j^{(\text{eff})})\mathbf{U}(j)], \tag{6.15}$$

and the trick is choosing the transformation that makes the cluster Hamiltonian best suited for its purpose. The most useful cluster model is one that represents the ground states as accurately as possible. To achieve this we can vary the parameters to make the minimum energy of the lowest APES of the cluster Hamiltonian, $\mathcal{H}_{\text{eff}}$, as low as possible. Introducing the rest of the Hamiltonian as a correction can only reduce the energy still further, so minimizing the cluster energy can only reduce these corrections. At strong coupling the energies of the minima in all the systems are some multiples of $k^2\hbar\omega$ below the uncoupled energy zero, so we set out to maximize $k_{\text{eff}}^2\omega_{\text{eff}}$. Some algebraic manipulation shows that this is done by choosing

$$s_1 = \frac{k_1}{\sqrt{k_1^2 + k_2^2}}, \qquad s_2 = \frac{k_2}{\sqrt{k_1^2 + k_2^2}}, \tag{6.16}$$

and with this choice we have

$$k_{\text{eff}}^2 = (k_1^2 + k_2^2) \quad \text{and} \quad \omega_{\text{eff}} = \frac{k_1^2 \omega_1 + k_2^2 \omega_2}{k_1^2 + k_2^2}. \tag{6.17}$$

6.1.1.2. Corrections to the Cluster Ground States

In order to assess the worth of this effective Hamiltonian as an approximation to the real thing, we must now proceed to make an estimate of how much it needs to be corrected.

The frequency of the remaining harmonic oscillator is

$$\omega_2' = (s_2^2\omega_1 + s_1^2\omega_2) = (k_2^2\omega_1 + k_1^2\omega_2)/k_{\text{eff}}^2, \tag{6.18}$$

and the interaction term can be written rather nicely as

$$\mathcal{H}_I = c[\alpha_j^{(2)\dagger}\eta_j + \eta_j^\dagger\alpha_j^{(2)}], \tag{6.19}$$

where

$$\eta_j = (\alpha_j^{(\text{eff})} + k_{\text{eff}}U(j)), \qquad \eta_j^\dagger = (\alpha_j^{(\text{eff})\dagger} + k_{\text{eff}}U(j)), \tag{6.20}$$

and

$$c = \hbar k_1 k_2(\omega_1 - \omega_2)/k_{\text{eff}}^2. \tag{6.21}$$

One point that can be seen at once is that if $\omega_1 = \omega_2$, then $c = 0$, and the cluster Hamiltonian is exact and needs no correction. This restates a rather obvious fact about taking linear combinations of coordinates, but this treatment does show immediately how the interaction strengths, k_1 and k_2, must be combined in that case. Another point is that c does not depend on the total strength of the coupling, only on the relative sizes of the k's and the degree of spread of the frequencies. We thus have a perturbation coefficient that does not blow up as the coupling strength increases, which is hopeful. The form of $\mathcal{H}_I$ is conveniently bilinear in operators that operate only within the eigenstates of $\mathcal{H}_{\text{eff}}$, the $\{\eta_j, \eta_j^\dagger\}$, and those that operate within the eigenstates of $\mathcal{H}_{ph'}$. This makes it easy to use as a perturbation if we can find the matrix elements of the $\{\eta_j, \eta_j^\dagger\}$. To do this we look at the commutators:

$$[\mathcal{H}_{\text{eff}}, \alpha_j^{(\text{eff})}] = -\hbar\omega_{\text{eff}}\eta_j, \qquad [\mathcal{H}_{\text{eff}}, \alpha_j^{(\text{eff})\dagger}] = +\hbar\omega_{\text{eff}}\eta_j^\dagger. \tag{6.22}$$

If the eigenstates of $\mathcal{H}_{\text{eff}}$ are written $|n\rangle$ with energy E_n, then the results of taking expectation values of the commutators (6.22) can be written

$$(E_{n'} - E_n)\langle n'|\alpha_j^{(\text{eff})}|n\rangle = -\hbar\omega_{\text{eff}}\langle n'|\eta_j|n\rangle \tag{6.23}$$

$$(E_{n'} - E_n)\langle n'|\alpha_j^{(\text{eff})\dagger}|n\rangle = +\hbar\omega_{\text{eff}}\langle n'|\eta_j^\dagger|n\rangle.$$

This shows immediately that if $E_{n'} = E_n$, the matrix elements of the η operators are zero, and, as the normal harmonic oscillator raising and lowering operators are non-diagonal also, this means that $\mathcal{H}_I$ will only enter as a perturbation at second order. Also, because the ground state of the system is also the ground state of the uncoupled oscillators, only the terms in $\eta_j \alpha_j^{(2)\dagger}$ operating on the ground state are effective in its second-order perturbation. It also follows from (6.20) that

$$\eta_j - \eta_j^\dagger = \alpha_j^{(\text{eff})} - \alpha_j^{(\text{eff})\dagger}, \tag{6.24}$$

so that taking linear combinations of (6.23) gives

$$(E_{n'} - E_n)\langle n'|\eta_j - \eta_j^\dagger|n\rangle = -\hbar\omega_{\text{eff}}\langle n'|\eta_j + \eta_j^\dagger|n\rangle \tag{6.25}$$

and

$$(E_{n'} - E_n)\langle n'|\alpha_j^{(\text{eff})} + \alpha_j^{(\text{eff})\dagger}|n\rangle = -\hbar\omega_{\text{eff}}\langle n'|\eta_j - \eta_j^\dagger|n\rangle. \tag{6.26}$$

Together these equations relate matrix elements of the η's to matrix elements of $\alpha_j^{(\text{eff})} + \alpha_j^{(\text{eff})\dagger}$, which is just a multiple of a coordinate variable, $q_j^{(\text{eff})}$, in the effective Hamiltonian. At strong coupling, the states most strongly connected by an operator that is just a function of the coordinates are those that overlap most in coordinate space, which for the ground state would be those on higher APESs localized near the potential minima. The reason for these vertical connections is the same as the reason for vertical transitions (see Section 4.6), namely, that at high quantum numbers the largest amplitude of a wave function occurs where the kinetic and potential energies are equal. The energies of the states vertically above the ground state depend on the differences of the eigenvalues of the matrix whose lowest eigenvalue gives the Jahn-Teller energy, and they are therefore multiples of $k_{\text{eff}}^2 \hbar\omega_{\text{eff}}$. These energy differences will give the perturbation denominators. The sum over states needed in a second-order perturbation expansion can be aided by the use of a sum rule that can be found by rewriting

$$\mathcal{H}_{\text{eff}} = \hbar\omega_{\text{eff}} \sum_j (\eta_j^\dagger \eta_j - k_{\text{eff}}^2 \mathbf{U}(j)^2) \tag{6.27}$$

and finding its expectation value in the ground state, to give

$$E_0 = \hbar\omega_{\text{eff}} \sum_j \langle 0|\eta_j^\dagger \eta_j|0\rangle - \kappa k_{\text{eff}}^2 \hbar\omega_{\text{eff}}. \tag{6.28}$$

The simplicity of the last term arises because $\sum_j \mathbf{U}(j)^2$ is a multiple of the unit matrix as long as the electronic basis belongs to a single representation of the appropriate symmetry group. This multiple is here put in as κ. This sum rule gives a value for the sum of the perturbation numerators

$$c^2 \sum_n \sum_j |\langle n|\eta_j|0\rangle|^2 = c^2 \sum_j \langle 0|\eta_j^\dagger \eta_j|0\rangle \tag{6.29}$$

by closure. The right-hand side of the equation is a multiple of $c^2 k_{\text{eff}}^2$, the exact multiplying factor depending on the Jahn-Teller system being considered. We now make the approximation that all the perturbation energy denominators are the same and correspond to the group of levels where the coupling is a maximum, that is, the splitting of the APESs at the location of the minima. This is a multiple of $k_{eff}^2 \hbar \omega_{eff}$, as remarked earlier. The energy of the uncoupled phonon is neglected in the energy denominators, being small compared with $k_{eff}^2 \hbar \omega_{eff}$. The result is a lowering of the energy by second-order perturbation that is a multiple of c^2 / ω_{eff}. Comparison of this with the expression for c (6.21) shows that this correction to the cluster energy depends only on the spread of the coupled mode frequencies and is of order $\hbar \omega_{eff}$. Since the cluster Jahn-Teller energy scales with $k_{eff}^2 \hbar \omega_{eff}$, this is a suitably small correction at strong coupling. A more refined treatment introducing second-order perturbations from other regions of the spectrum can find more corrections, but they are all of order $\hbar \omega_{eff} / k_{eff}^2$ or smaller. These corrections can be expected to alter the relative energies of any pseudo-rotational levels, but on a larger scale their effect is small.

6.1.1.3. The N-Mode Cluster Hamiltonian

The next question to be addressed is whether this cluster approximation gets worse if the number of coupled modes increases. The answer is that it does not. The algebra to show this can be done quite generally for an arbitrary number of coupled modes, but as it is more complicated than the two-mode case and since it can be found in O'Brien (1972), we shall simply quote the result here. As for the two-mode case, the Hamiltonian for N coupled modes,

$$\mathcal{H}_N = \sum_{i=1}^{N} \hbar \omega_i \sum_j [a_j^{(i)\dagger} a_j^{(i)} + k_i (a_j^{(i)\dagger} + a_j^{(i)}) U(j)], \qquad (6.30)$$

can be transformed into the form

$$\mathcal{H} = \mathcal{H}_{eff} + \mathcal{H}_{ph'} + \mathcal{H}_I \qquad (6.31)$$

with

$$\mathcal{H}_{eff} = \hbar \sum_j [\omega_{eff} \alpha_j^{(eff)\dagger} \alpha_j^{(eff)} \qquad (6.32)$$
$$+ k_{eff} \omega_{eff} (\alpha_j^{(eff)} + \alpha_j^{(eff)\dagger}) U(j)]$$

$$\mathcal{H}_{ph'} = \hbar \sum_{i=2}^{N} \sum_j \omega_i' \alpha_j^{(i)\dagger} \alpha_j^{(i)}$$

$$\mathcal{H}_I = \hbar \sum_{i=2}^{N} \sum_j c_i [\alpha^{(i)\dagger} \eta_j + \alpha^{(i)} \eta_j^\dagger],$$

where

$$k_{\text{eff}}^2 = \sum_{i=1}^{N} k_i^2 \quad \text{and} \quad k_{\text{eff}}^2 \omega_{\text{eff}} = \sum_{i=1}^{N} k_i^2 \omega_i. \tag{6.33}$$

The ω_i' are a new set of frequencies for the remaining $N - 1$ oscillators, and they are distributed in such a way that there is one of the ω' frequencies between every adjacent pair of the original frequencies. Individual values of the c_i cannot be found easily, but the quantity required for the second-order perturbation can be found. It is

$$\sum_{i=2}^{N} c_i^2 = \hbar^2(\langle \omega^2 \rangle - \langle \omega \rangle^2), \tag{6.34}$$

with

$$\langle \omega^2 \rangle = \sum_{i=1}^{N} k_i^2 \omega_i^2 \quad \text{and} \quad \langle \omega \rangle = \omega_{\text{eff}}. \tag{6.35}$$

If the coupling strengths are regarded as a distribution over the frequencies, then $\sum_i c_i^2$ is the mean square width of this distribution. Thus we see that the correction to the cluster energy in the ground state still depends only on the spread of the frequency distribution and not on the coupling strength, and this mean square width should be some fraction of ω_{eff}^2. The aim of the foregoing discussion is not to produce detailed forms for the corrections to the recommended cluster, but rather to estimate how good the cluster will be as a model for the ground state of a multimode Jahn-Teller system. So far we have shown that it is rather good at strong coupling. This is illustrated in Figure 6.1, where the energy of the lowest state of $p \otimes h$ is plotted against k_{eff}^2. To calculate the extra multi-mode corrections, an array of eight frequencies running from ω to 5ω was chosen for this figure, with equal coupling to every mode. This is meant to simulate roughly the coupling of the t_{1u} orbital to the eight h_g modes of C_{60} (Table 6.2). This is a relatively wide band of coupled modes, but the corrections remain quite small. At weak coupling this cluster model is also a good approximation to the ground state, as Figure 6.1 also illustrates. To see this, consider first the cluster Hamiltonian $\mathcal{H}_{\text{eff}}$ (6.15). Within the uncoupled states the Jahn-Teller part has no diagonal matrix elements because it only contains raising and lowering operators. It first appears in second-order perturbation connecting the first excited oscillator state to the ground state. Hence all the energy denominators are equal to $\hbar\omega_{\text{eff}}$, and the matrix elements of the η's are unity. The matrix of the second-order perturbation in the ground states is thus

$$V_{\text{eff}}^{(2)} = k_{\text{eff}}^2 \omega_{\text{eff}} \sum_j \mathbf{U}(j)^2. \tag{6.36}$$

The N-mode Hamiltonian (6.30) also has no diagonal matrix elements. Its ground state is the product of all the oscillator ground states multiplied by the electronic base. It connects the ground state with states that have a single

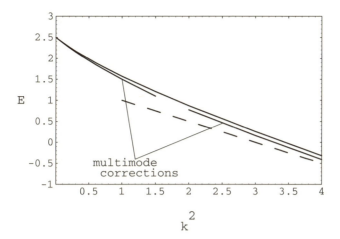

Figure 6.1. Calculated energy for the ground state of $p \otimes h$, showing the multi-mode corrections expected for C_{60}^- at large and small coupling strength. The corrections are modeled as eight frequencies in the range $(\omega, 5\omega)$ with equal coupling to each. The dashed line shows the asymptotic energy.

excitation of one mode of vibration (because the Hamiltonian contains only terms linear in the a's), so the complete matrix of second-order perturbation in the ground states is

$$V^{(2)} = \sum_{i=1}^{N} k_i^2 \omega_i \sum_j \mathbf{U}(j)^2, \qquad (6.37)$$

and clearly this is exactly given by V_{eff}. Thus the cluster Hamiltonian is exact to second order in perturbation. It requires correction at fourth order. This correction can be found for each Jahn-Teller system in terms of sums over modes, but it is not expressible simply in k_{eff} and ω_{eff}. This fourth-order correction at weak coupling is included in Figure 6.1. This leaves us with the problem of intermediate coupling strengths. As usual this would require numerical work, which has not been done in any icosahedral system. However it has been done by Evangelou et al. (1980) in the much simpler case of an E state in cubic symmetry coupled to two ϵ modes. There it was found that the correction to the ground state varied smoothly between strong and weak coupling, with the largest correction occuring at about $k_{\mathrm{eff}} = 1$.

6.1.1.4. The Ham Factors

If we use $\mathcal{H}_{\mathrm{eff}}$ to find ground state energies, we must hope to be able to use it for the Ham factors, which encapsulate the measurable properties of the ground state eigenvectors. Sadly no general theory has been produced that links all the Jahn-Teller systems, so we are forced back entirely on the analogy with $E \otimes \epsilon$. Here the numerical work of Evangelou et al. (1980) found that the Ham factors were given correctly by $\mathcal{H}_{\text{eff}}$ at weak and strong coupling, and the corrections needed at intermediate coupling strength remained small. Analytic work confirmed the numerical results in the two extreme regimes.

6.1.2 A Cluster for Optical Absorption

The cluster Hamiltonian that does best for optical absorption bands in the multi-mode Hamiltonian (6.30) has parameters for $\mathcal{H}_{\text{eff}}$ given by the relations

$$k_{\text{eff}}^2 \omega_{\text{eff}}^2 = \sum_{i=1}^{N} k_i^2 \omega_i^2 \tag{6.38}$$

$$k_{\text{eff}}^2 \omega_{\text{eff}} = \sum_{i=1}^{N} k_i^2 \omega_i,$$

which give different values of k_{eff} and ω_{eff} from (6.33). The situation for which this cluster works occurs when the Jahn-Teller interaction is in an excited state and optical transitions are taking place to that state from an undistorted ground state. The choice of cluster is made by matching moments of the band at strong coupling. The use of moments to characterize the properties of an absorption band is a well-established experimental technique (e.g. Henry et al., 1965), and it rests on the fact that the moments of a band are given by the relation

$$\mu_n = \langle 0 | \mathcal{H}^n | 0 \rangle, \tag{6.39}$$

where $\mathcal{H}$ is the interaction producing the band width, and μ_n is the nth moment of the band about the uncoupled excited state $|0\rangle$. In our case $\mathcal{H}$ is the multimode Hamiltonian $\mathcal{H}_N$ (6.30). We can write

$$\mathcal{H}_N = \mathcal{H}_{ph} + \mathcal{H}_{JT}, \tag{6.40}$$

where

$$\mathcal{H}_{ph} = \sum_{i=1}^{N} \hbar \omega_i \sum_j a_j^{(i)\dagger} a_j^{(i)} \tag{6.41}$$

$$\mathcal{H}_{JT} = \sum_{i=1}^{N} \hbar \omega_i k_i \sum_j (a_j^{(i)\dagger} + a_j^{(i)}) \mathbf{U}(j),$$

so that

$$\mathcal{H}^n = \mathcal{H}_{JT}^n + (\mathcal{H}_{JT}^{(n-1)} \mathcal{H}_{ph} + \mathcal{H}_{JT}^{(n-2)} \mathcal{H}_{ph} \mathcal{H}_{JT} + \cdots + \mathcal{H}_{ph} \mathcal{H}_{JT}^{(n-1)}) + \cdots \tag{6.42}$$

and this is the start of an expansion in successively lower powers of the k_i's. Now $\mathcal{H}_{JT}$ is linear in the raising and lowering operators, so only even powers of it have nonzero diagonal matrix elements, unlike $\mathcal{H}_{ph}$ which contains products such as $a^\dagger a$. Consequently the first term in the expansion is nonzero only when n is even, and the second term only when n is odd, so the two terms together give the highest order term in both even and odd moments. The choice of cluster parameters, (6.38) and (6.39), is the one that matches the term of largest order in the k's in the even and odd moments respectively, so that for every n

$$\langle 0|\mathcal{H}_{\text{eff}}^{(n)}|0\rangle = \langle 0|\mathcal{H}_N^{(n)}|0\rangle \tag{6.43}$$

to the highest order in k_{eff}. Work showing that this is so for the cluster parameters (6.38) is given by Fletcher et al. (1979). These authors have also applied the test of the numerical solution of a cubic E state coupled to two ϵ modes and find that it gives a good representation of the smoothed band shape even when k_{eff} is as small as 1. This is a rather strange approximation, because it replaces the multi-mode spectrum with its many excitations with a single-mode spectrum that necessarily has many fewer widely spaced lines. The matching only occurs if the discrete lines are smeared out, as they would be in a solid or a very large molecule, by coupling to symmetric modes of vibration. In the numerical comparisons this coupling is simulated by convolution with a Gaussian band, with the result described above.

6.2 ELECTRON SPIN RESONANCE

As remarked in Chapter 1, the first experimental evidence of a Jahn-Teller effect was in work in electron spin resonance (esr) by Bleaney and Bowers (1952). They were studying the spectrum of the $Cu^{++}[H_2O]_6$ complex, and they observed a spectrum that went from that of a set of distorted sites at low temperatures to being isotropic at higher temperatures. This was interpreted as being due to an $E \otimes \epsilon$ Jahn-Teller coupling with warping that produced a set of wells (see Appendix C). The low temperature spectrum then corresponds to the system being trapped in one or other of the wells, each well giving a different low symmetry spectrum, while the high temperature spectrum is an average of the single well spectra produced by thermally assisted tunneling. It was later pointed out by Ham (1968) that the random strains in the crystal were crucial in producing the distorted spectra. Another manifestation of the Jahn-Teller effect, also in Cu^{++} (in MgO), was first reported by Coffman (1965). This spectrum had a different angular variation from anything seen before, and it was interpreted as being due to a fairly strong Jahn-Teller effect with weak warping, so that the system tunnels freely from well to well and the ground vibronic state is of the same symmetry type and degeneracy as the uncoupled electronic state. We shall see that both these types of esr can in principle be

TABLE 6.1
Products of Irreps with Γ_6,
Showing States Resulting
from Singly Occupied
Orbitals

$A \otimes \Gamma_6$	$=$	Γ_6
$T_1 \otimes \Gamma_6$	$=$	$\Gamma_6 \oplus \Gamma_8$
$T_2 \otimes \Gamma_6$	$=$	Γ_9
$G \otimes \Gamma_6$	$=$	$\Gamma_7 \oplus \Gamma_9$
$H \otimes \Gamma_6$	$=$	$\Gamma_8 \oplus \Gamma_9$

found in icosahedral symmetry, with the latter associated with T_2 orbitals. More often the effect of the Jahn-Teller coupling on esr spectra is less dramatic. The spectrum may have a lower symmetry than is expected from the symmetry of the site. If this occurs as a result of Jahn-Teller coupling, then random strains must be taken into account too, but symmetry lowering may be associated with local strains from other causes. The magnitude of esr parameters may be altered by the various Ham factors, but unless the Jahn-Teller quenching is very marked, it will be difficult to disentangle this effect from everything else. Experience of esr in cubic symmetry has been so very much more extensive than in icosahedral symmetry that it is not surprising that we have to look to cubic symmetry for unambiguous examples of the Jahn-Teller effect. In particular the best information is obtained from single crystal samples where the position of the magnetic field relative to the symmetry axes can be controlled, and this is difficult to achieve for icosahedral complexes, since such molecules are not easily crystallized. Spectra from powder specimens are necessarily averaged over all directions, and much information is lost in consequence. However in the theoretical sections that follow, we shall list what might be found under favorable circumstances.

6.2.1 Jahn-Teller Interactions in the Spin Representations of the Group I

The double-valued, or spin, representations of I were introduced in Chapter 2, and there are a few more remarks that should be made about them before we discuss spin resonance. Of the four spin representations, only two, Γ_8 and Γ_9, are subject to a Jahn-Teller interaction, and the active modes for each are derived from the Kronecker products shown in Table 2.4. The complete table of products of the irreps of the double group is shown in Appendix D (D.3). Table 6.1 is extracted from it to show the effect of coupling electronic states of each symmetry type to a single electron spin (an electron spin transforms like Γ_6). One use of this table is to pick out possible splittings. It is the fact that a T_2 orbital state remains unsplit in this symmetry when coupled to a spin that makes it a special case for spin resonance. Another important consideration for spin resonance deals with which of the irreps can have nonzero angular momentum. The angular momentum operator transforms as the T_1 irrep of I, and in Table D.3, T_1 appears in the square of every irrep except T_2 and Γ_7, so only these two irreps are intrinsically nonmagnetic.

6.2.1.1. $\Gamma_8 \otimes h$

The $\Gamma_8 \otimes h$ system is an interesting one because its Hamiltonian has accidental SO(5) symmetry. This means that at strong coupling it has a large density of low-lying pseudo-rotational levels (rotations in five-dimensional space), and it has only two different Ham factors, which can be calculated numerically at all coupling strengths. $K(H)$ goes down to $1/5$ at strong coupling, while $K(T_1)=K(T_2)=K(G)$ goes from 1 down to $3/5$. These calculations have been made for the identical system $\Gamma_8 \otimes (t_2 \oplus e)_{eq}$ in cubic symmetry by Pooler and O'Brien (1977).

6.2.1.2. $\Gamma_9 \otimes (g \oplus h)$

Like $H \otimes H$, $\Gamma_9 \otimes \Gamma_9$ contains H twice as a Jahn-Teller active mode, so that the coupling of any h mode must be expressed as a linear combination of two independent matrices. Thus the $\Gamma_9 \otimes (g \oplus h)$ system will be very complicated, with many possibilities inherent in the various possible relative coupling strengths. Considering that Γ_9 can arise by the coupling $H \otimes \Gamma_6$, it is clear that all the possible extra symmetry invariants found in H coupling and discussed in Chapter 5 must reappear in $\Gamma_9 \otimes (g \oplus h)$.

6.2.2 The Spin Hamiltonian

The standard way of relating the results of resonance experiments to theory is through use of the spin Hamiltonian. This has been extensively used and studied, and Abragam and Bleaney (1970) have published one of many texts expounding it. In the following discussion we omit reference to the effects of nuclear spin and quadrupole moments for simplicity. In this method the actual Hamiltonian is replaced by a Hamiltonian that operates within a set of spin states, where the spin is chosen so that it has the same degeneracy as the actual set of states among which the resonance is taking place. The parameters in the spin Hamiltonian can be chosen to be the largest set allowable under the symmetry of the Hamiltonian and no more. Then both theory and experiment try to find the values of this set of parameters, so that the spin Hamiltonian is the place where theory and experiment meet. The reason that this procedure works is that the energies involved in the resonance process are normally quite small, so that perturbation within a set of degenerate or nearly degenerate states is all that needs to be considered, and the magnetic field is only taken to first order. The simplest case is a two-level system, normally a Kramers doublet, which is represented by the spin Hamiltonian for $S = 1/2$. In low symmetry the most general spin Hamiltonian is

$$H_S = \mu_B \mathbf{B} \cdot \overline{\mathbf{g}} \cdot \mathbf{S}, \tag{6.44}$$

where $\overline{\mathbf{g}}$ is a tensor, $\mathbf{B}$ is the magnetic field, and μ_B the Bohr magneton. In icosahedral symmetry the irrep Γ_6 can be $J = 1/2$, which has an isotropic interaction with a field, so all occurrences of Γ_6 must be isotropic, and we have instead

$$H_S = g\mu_B \mathbf{B} \cdot \mathbf{S}, \tag{6.45}$$

where g, the g-factor, is a single constant. Table D.3 tells us that Γ_7 is nonmagnetic, since its Kronecker product does not contain T_1, which represents the angular momentum operator: its g-factor is thus 0. Next consider a three-level system, which must have a spin-one Hamiltonian. This is traditionally written

$$H_S = \mu_B \mathbf{B} \cdot \overline{\mathbf{g}} \cdot \mathbf{S} + D \left(S_z^2 - \frac{1}{3} S(S+1) \right) + E(S_x^2 - S_y^2), \tag{6.46}$$

where the terms in D and E introduce a second-order tensor operator referred to its principal axes and adjusted to have zero trace for convenience. This tensor is not an icosahedral invariant, so its presence implies lower symmetry, as does an anisotropic $\overline{\mathbf{g}}$. The triangle rule for the addition of angular momenta shows that with $S = 1$, no higher order tensors are needed, but as S increases so does the number of tensors (and independent parameters) that may have to be introduced.

6.2.3 The g Factors

The g factors have to represent the effect of the magnetic field energy operator, $\mu_B (\mathbf{L} + 2\mathbf{S}) \cdot \mathbf{B}$ within the set of states in which the resonance is being observed. There are normally two main regimes to be considered, according to whether the coupling of the orbit and spin angular momenta is strong or weak compared with the effect of the environment, which here has icosahedral symmetry. The strong case usually requires substantial spin-orbit coupling, and is seen in heavy, well shielded ions such as the rare earths. The weak case is to be expected in transition metal ions and in large molecules like C_{60}.

6.2.3.1. The g Factors with Strong Spin-Orbit Coupling

This is the rare-earth-like situation, which might be found if a magnetic ion was trapped inside an icosahedral cage. The ground state of the free ion would be characterized by a value of J, and the magnetic energy would be given by $\mu_B g_J \mathbf{J} \cdot \mathbf{S}$, where g_J is the appropriate Landé g factor. This state would be split into icosahedral irreps according to one of the lists in Table 2.2 or Table 2.3. For $J \leq 5/2$ there is no splitting, the g factor is isotropic, and in the absence of Jahn-Teller coupling would be equal to the g_J. As the Jahn-Teller coupling is turned on, this g factor would be quenched by multiplication by the Ham factor $K(T_1)$. If the angular momentum was higher and the icosahedral splitting produced one of the irreps T_1, H, Γ_7, Γ_8, or Γ_9, then the g factor would still be isotropic, its value would be a multiple of g_J, and it would also be quenched by $K(T_1)$. As already remarked, T_2 and Γ_7 are nonmagnetic and would have $g = 0$. This leaves only G for special consideration, except for A and Γ_6, which have no Jahn-Teller coupling. According to the rules, the fourfold degenerate G state should be represented by a spin Hamiltonian with $S = 3/2$, but the form of the angular momentum matrices fits exactly with two Γ_6 states with the same g factor, and this would probably be the most useful way to represent it. Unlike the g factor in Γ_6, however, this g factor would be quenched by $K(T_1)$. If the Jahn-Teller interaction were strong in these cases, so that the minima in the lowest APES were deep, then the esr spectrum would correspond to static distortions at the minima, and would be characteristic of the distorted symmetry. The static distortion would arise as the result of the combination of the Jahn-Teller interaction and local strains or crystal fields. Under the Jahn-Teller interaction alone, there are a number of wells of equal energy, so that the vibronic ground states are a linear combination of states in all the wells. However, if the barriers between the wells are not too low, a small low symmetry term in the potential is all that is needed to localize the wave function in one well that drops lower than the others. Such a low symmetry term can easily arise as a result of strain fields, and the prevalence of strains ensures that spectra will more often be of a superposition of distorted sites than of quantum-mechanical averages.

6.2.3.2. The g Factors for Spin Doublet States

We start the discussion of spin doublet states with the case when the Jahn-Teller interaction is strong enough to override any spin-orbit coupling, so that the ground state of the ion under the Jahn-Teller interaction is found first and the spin is coupled in afterwards. The ground electronic state is then an orbital singlet, so the orbital angular momentum is 0 and all the magnetism is carried by the electron spin, which would produce an isotropic g factor of 2 in a $S = 1/2$ spin Hamiltonian. This is not the whole story, however. If there is a static distortion of symmetry D_{3d} or D_{5d} then the most general form of the spin Hamiltonian has an anisotropic, axially symmetric $\bar{\mathbf{g}}$ tensor, which is conventionally written

$$H_S = \mu_B(g_\parallel B_z S_z + g_\perp B_x S_x + g_\perp B_y S_y), \tag{6.47}$$

with the z direction along the axis of symmetry. With such a Hamiltonian and **B** at an angle θ to the z direction, there will be a single line at an energy $\mu_B B \sqrt{(g_\parallel^2 \cos^2\theta + g_\perp^2 \sin^2\theta)}$. If the distortion symmetry is lower, the $\overline{\mathbf{g}}$ tensor can have three different principal values, and the spin Hamiltonian is conventionally written

$$H_S = \mu_B(g_x B_x S_x + g_y B_y S_y + g_z B_z S_z). \tag{6.48}$$

These departures of g from the spin-only value are commonly found in the transition metal ions, and they are caused by admixtures of excited states via the spin-orbit operator.

6.2.3.3. $^2T_1$ with Static Distortion

Consider a $^2T_{1u}$ or ^2p state. Under a D_{3d} or D_{5d} Jahn-Teller distortion, there is a p_z ground state (z being the symmetry axis) and degenerate excited states p_x and p_y. The spin-orbit coupling, $\lambda \mathbf{L} \cdot \mathbf{S}$, mixes excited states into the ground state with an admixture coefficient of order λ/Δ, where Δ is the Jahn-Teller splitting, and this admixture allows the orbital part of the magnetic moment operator $\mu_B \mathbf{B}.\mathbf{L}$ to be nonzero. This process has been usefully formalized in the following way (see Abragam and Bleaney, 1970, Chapter 19): in second-order perturbation theory the effect of the two operators together can be expressed as

$$-\sum(\mu_B \mathbf{B} \cdot \mathbf{L} + \lambda \mathbf{L} \cdot \mathbf{S})^2/\Delta, \tag{6.49}$$

where the sum is over excited states. For the g factors, we are only interested in bilinear terms in **B** and **S**, so the operator giving the corrections can be written

$$-\mu_B \frac{\lambda}{\Delta} \mathbf{B} \cdot \overline{\Lambda} \cdot \mathbf{S}, \tag{6.50}$$

where $\overline{\Lambda}$ is the symmetric tensor given by

$$\Lambda_{ij} = \langle 0|L_{ij}^{(2)}|0\rangle, \quad \text{with} \quad L_{ij}^{(2)} = L_i L_j + L_j L_i, \tag{6.51}$$

$|0\rangle$ being the orbital ground state. This is correct if all the perturbation states are at the same energy, Δ. In the $^2T_{1u}$ or ^2p state, this gives no correction to $g_\parallel$ but does provide a correction to $g_\perp$ of $-2\lambda/\Delta$. With well-localized Jahn-Teller minima, we should thus expect to see a superposition of spectra corresponding to the spin Hamiltonian (6.47) and the g factors above, each with its z-axis along one of the Jahn-Teller distortion axes. The above discussion started off in terms of localization at the Jahn-Teller minima produced by warping, but at any point on the spherical minimum surface, the p state will be split with p_z as the ground state and p_x and p_y above it at energy Δ, where z is along the distortion axis. Consequently the $\overline{\mathbf{g}}$ tensor will have axial symmetry everywhere on the spherical surface, so the effect of strain fields picking out points on the spherical surface will also look like a powder spectrum.

6.2.3.4. $^2T_1$ States with Weak Warping

When the localization of the minima is not so strong, because the warping is weak, we must look for the correction to g for a degenerate T_1 ground state. That is to say, we are looking at a perturbation matrix in the form

$$-\mu_B \frac{\lambda}{\Delta} \langle T_1 | B_i L_{ij}^{(2)} S_j | T_1 \rangle. \tag{6.52}$$

Since Δ is the Jahn-Teller splitting, this treatment by second-order perturbation implies strong Jahn-Teller coupling. The main part of the spin Hamiltonian on which this is a perturbation is the interaction of the field with the electron spin, $2\mu_B \mathbf{B} \cdot \mathbf{S}$, which lines up the spin along the field, so that $\mathbf{S}$ is parallel to $\mathbf{B}$. Thus if the direction cosines of $\mathbf{B}$ are ℓ_x, ℓ_y, ℓ_z ($\ell_x^2 + \ell_y^2 + \ell_z^2 = 1$), then $B_i S_j$ in the perturbation can be replaced by $B S_\zeta \ell_i \ell_j$, where S_ζ is the component of the spin along the field, $\pm 1/2$. The perturbation then becomes

$$-\mu_B \frac{\lambda}{\Delta} B S_\zeta \ell_i \ell_j \langle T_1 | L_{ij}^{(2)} | T_1 \rangle, \qquad S_\zeta = \pm 1/2. \tag{6.53}$$

To use this in a general way, we can start by noticing that, as a second-order symmetric tensor, the operator $L_{ij}^{(2)}$ can be split up into two parts that transform as A and H under the icosahedral group. The A part is just $\frac{2}{3}(L_x^2 + L_y^2 + L_z^2)$; it is diagonal in any state and adds an isotropic correction to the isotropic spin g factor. The tensor $\ell_i \ell_j$ can be similarly decomposed. The matrix of the H part of the perturbation in the T_1 states hence takes the same form as the familiar matrix (3.2) with the Q_i's replaced by the equivalent components of the $\ell_i \ell_j$ tensor:

$$\begin{bmatrix} \frac{1}{6}(3\ell_z^2 - 1) - \frac{1}{2}(\ell_x^2 - \ell_y^2) & -\ell_x \ell_y & -\ell_x \ell_z \\ -\ell_x \ell_y & \frac{1}{6}(3\ell_z^2 - 1) + \frac{1}{2}(\ell_x^2 - \ell_y^2) & -\ell_y \ell_z \\ -\ell_x \ell_z & -\ell_y \ell_z & -\frac{1}{3}(3\ell_z^2 - 1) \end{bmatrix} \tag{6.54}$$

The eigenvalues of this matrix, as we saw in Section 3.2.1, are independent of direction in the (ℓ_x, ℓ_y, ℓ_z) space, and consequently independent of the direction of $\mathbf{B}$. At all directions of the field there will be one transition at g_1 and two at g_2, and the whole spectrum will be isotropic. These perturbative corrections to the g factor will be quenched by $K(H)$, which tends to 2/5 at strong coupling. The two g factors are given by

$$\begin{bmatrix} g_1 \\ g_2 \end{bmatrix} = \left(2 - \frac{4\lambda}{3\Delta}\right) \begin{bmatrix} 1 \\ 1 \end{bmatrix} + \frac{2\lambda}{3\Delta} K(H) \begin{bmatrix} 2 \\ -1 \end{bmatrix}. \tag{6.55}$$

This highly isotropic spectrum is what would be expected in exact icosahedral symmetry; however, care should be taken: experiments on esr in sites that are nominally of cubic symmetry show that perturbations by local strain fields are almost always sufficient to reduce the symmetry enough to alter the appearance of the spectrum, and the same should be expected here. In any case the icosahedral symmetry cannot extend out into a crystal lattice, so there will be crystal fields that will play the role that strains do in cubic symmetry. There are two possible strain or crystal field effects. If we assume that the Jahn-Teller coupling remains large, but that there is little warping, then the effect of a strain field will be to lower the energy of one point on the spherical minimum relative to the rest. The result at that site will be a $\bar{\mathbf{g}}$ tensor with its principal axis along the direction selected by the strain. (The strain tensor also transforms as H.) If the strains are random, then the resulting spectrum will be like that of a powder, an average of axial spectra over all directions of the field. If the warping potential is large enough to select out either the D_{5d} or D_{3d} minima, then the effect of strains will be to make one or other of these minima lower than the others, and the spectrum will be the superposition of distorted spectra described in Section 6.2.3.3.

6.2.3.5. $^2T_2$ States with Weak Warping

We can now proceed to consider the $^2T_2$ state, which differs from $^2T_1$ both in having no orbital angular momentum, and in having a different form of interaction with an H-type operator. If the g factors are to differ from 2.0023, the spin g factor, there must be some off-diagonal angular momentum matrix elements, which must all be to a set of degenerate levels for a similar treatment to hold. This last condition would be satisfied if, for instance, the T_2 state derived from an $L = 3$ state in an icosahedral field. We assume this to be the case and use the theory developed for $^2T_1$ in a suitably modified form. As was the case for T_1, we look for a perturbation in the form

$$- \mu_B \frac{\lambda}{\Delta} B S_\zeta \ell_i \ell_j \langle T_2 | L_{ij}^{(2)} | T_2 \rangle, \qquad S_\zeta = \pm 1/2. \qquad (6.56)$$

However, now the energy difference Δ is to a set of states of different symmetry; it is not the Jahn-Teller splitting. The matrix of the H-type operator in T_2 is different from that in T_1, as is given explicitly in Appendix E (E.12). Putting in the appropriate forms of the components of $\ell_i \ell_j$, we get the matrix

$$\begin{bmatrix} \frac{1}{6}(2\ell_z^2 - \ell_x^2 - \ell_y^2) - \ell_x \ell_z & \ell_y \ell_z & \frac{1}{2}(\ell_x^2 - \ell_y^2) \\ \ell_y \ell_z & \frac{1}{6}(2\ell_z^2 - \ell_x^2 - \ell_y^2) + \ell_x \ell_z & \ell_x \ell_y \\ \frac{1}{2}(\ell_x^2 - \ell_y^2) & \ell_x \ell_y & -\frac{1}{3}(2\ell_z^2 - \ell_x^2 - \ell_y^2) \end{bmatrix} \qquad (6.57)$$

for the H part of $\ell_i \ell_j$. The eigenvalues of this matrix are the roots of the following cubic equation:

$$\mathcal{E}^3 - \frac{1}{3}\mathcal{E} + \frac{1}{168} V_{icos}(\ell_x, \ell_y, \ell_z) + \frac{4}{189} = 0, \qquad (6.58)$$

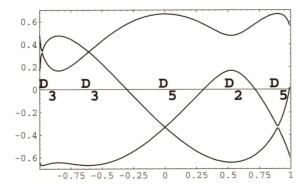

Figure 6.2. Eigenvalues of the matrix (6.57) on a section $\ell_y = 0$, plotted against ℓ_x. The points of D_{5d}, D_{3d}, and D_{2h} symmetry on the section are shown.

where V_{icos} is the icosahedral invariant that was given as a function of (x, y, z) in (3.14). The three g factors are given by expressions of the form

$$g_i = 2 + \frac{\lambda}{\Delta}(c_A + c_H K(H)\mathcal{E}_i), \tag{6.59}$$

where the $\mathcal{E}_i$ are the roots of the matrix (6.58), and c_A and c_H are numbers that depend on the provenance of the particular set of T_2 orbitals. These roots have a complicated angular variation: in a general direction there are three distinct roots, but two of them coalesce when **B** is along D_{3d} or D_{5d} directions. This can be seen in Figure 6.2. Because of the absence of angular momentum within T_2, this perturbation treatment can be applied at any coupling strength. The coupling strength only enters through $K(H)$. We thus have here the prediction of a spectrum with an extremely exotic angular variation, analogous to that of ^2E in cubic symmetry. However, as remarked before, experience with the cubic ^2E state suggests that this spectrum is often likely to be overridden by local strains.

6.2.3.6. $^2T_2$ with Static Distortion

With warping and/or strain active, so that a particular electronic T_2 ground state has been selected, the formulae to be used for the $\bar{\mathbf{g}}$ tensor are those given in (6.50) and (6.51). This is assuming, as in Section 6.2.3.5, that the T_2 state in question is connected via the angular momentum operator to a set of states at energy Δ, as would be the case for an $L = 3$ state in an icosahedral field.[1] To evaluate $\langle 0|L_{ij}^{(2)}|0\rangle$ (6.51), the L_{ij} operator is again split into an H and A part.

---

[1] This Δ cannot be the Jahn-Teller splitting in this case, because a T_2 state has no angular momentum.

The A operator gives a uniform g-shift, and it is the H operator that determines the principal axes of the $\overline{\mathbf{g}}$ tensor. Now if the ground state lies somewhere on the spherical Jahn-Teller minimum, then the Öpik and Pryce equation (4.8) tells us that in the electronic state on the lowest APES, the expectation value of each H-type operator is proportional to the value of the equivalent h normal coordinate at that point. This means that the H part of the tensor $\overline{\Lambda}$ (6.50) can be directly replaced by a matrix proportional to (6.57), with ℓ_x, ℓ_y, ℓ_z the coefficients of the three components of T_2 in the ground state. The eigenvalues of (6.57) are thus proportional to the principal values of the anisotropic part of $\overline{\Lambda}$ and hence to the anisotropy in the principal values of $\overline{\mathbf{g}}$. Accordingly the three principal values of the $\overline{\mathbf{g}}$ tensor are given by (6.59), but with $K(H) = 1$, and in general there are three different values. This is in contrast to $^2T_1$ states, which, as we saw earlier, will have a $\overline{\mathbf{g}}$ tensor of axial symmetry. The directions of the axes of the $\overline{\mathbf{g}}$ tensor do not come directly out of this calculation because (ℓ_x, ℓ_y, ℓ_z) are not directions in space. They could be calculated with some trouble, but if the distortion is of D_{5d} or D_{3d} type, the symmetry requires the $\overline{\mathbf{g}}$ tensor to be axial, with its axis along the particular fivefold or threefold axis. The above calculation started from the assumption that the electronic state satisfied the Öpik and Pryce conditions, but as it is also a perfectly general linear combination of the three T_2 bases, the g factors can always be found in this way.

6.2.4 The Spin Hamiltonian for Spin Triplet States

Here we address ourselves to the origin of the terms in D and E that appear in the spin Hamiltonian (6.46). There are two main sources, spin-orbit and spin-spin coupling. We start with the effect of spin-orbit coupling because it follows from the treatment of the g factors already given. Returning to equation (6.49), all that has to be done is to pick out the term

$$-\sum (\lambda \mathbf{L} \cdot \mathbf{S})^2 / \Delta \qquad (6.60)$$

and treat it as a tensor product in much the same way as before.

6.2.4.1. The D Tensor in $T \otimes h$ *with Static Distortion*

Within the spin states the second-order tensor is represented by the D and E terms in the spin Hamiltonian. The form of the terms in the spin Hamiltonian are chosen to have trace zero, so, unlike the $\overline{\mathbf{g}}$ tensor, only the H part of the operator need be considered, and the operator can be used in the form

$$-\frac{\lambda^2}{\Delta} \sum_{ij} S_{ij}^{(2)} L_{ij}^{(2)}, \qquad (6.61)$$

where $S_{ij}^{(2)}$ and $L_{ij}^{(2)}$ are assumed to be quadratic H-type operators. The standard form of the spin Hamiltonian is also chosen so that $E = 0$ corresponds to axial symmetry. The treatment of this operator follows very closely that for the $\bar{\mathbf{g}}$ tensor. The components of $L_{ij}^{(2)}$ operating in the electronic state give a set of values of Λ_{ij} as before, which in turn act as coefficients for the components of $S_{ij}^{(2)}$. The whole thing then operates within the set of spin states corresponding to $S = 1$. For T_2 the principal values of the D tensor come out as the roots of (6.57), but with (S_x, S_y, S_z) replacing (ℓ_x, ℓ_y, ℓ_z). Axial symmetry, corresponding to $E = 0$, only appears when the external symmetry requires it. For T_1 the D tensor is always axial and D is constant, similarly to the $\bar{\mathbf{g}}$ tensor.

6.2.4.2. T ⊗ h with Weak Warping

To handle degenerate states, we return to the argument that produced the perturbation in the form (6.53). We can write the perturbation for a spin triplet state in T_1 or T_2 in the form

$$-\left(\mu_B \frac{\lambda}{\Delta} B S_\zeta + \frac{\lambda^2}{\Delta} S_\zeta^2\right) \ell_i \ell_j \langle T|L_{ij}^{(2)}|T\rangle, \qquad S_\zeta = 0, \pm 1. \qquad (6.62)$$

With this expression the spectra for $^3T_1$ and $^3T_2$ can be predicted exactly as for the doublet states. Depending on the direction of the field, there will be two or three sets of allowed transitions. For each orbital state the two $\Delta S_\zeta = \pm 1$ transitions are allowed. These have energy $g\mu_B B$, where the g's are those calculated for the spin doublet states. The extra term in S_ζ^2 splits these transitions by an amount proportional to the appropriate eigenvalue of the $\langle T|\ell_i\ell_j L_{ij}^{(2)}|T\rangle$ matrix. Hence for $^3T_1$ the spherical symmetry is preserved, while the angular variation for $^3T_2$ continues to depend on the roots of the equation (6.58).

6.2.4.3. The Spin-Spin Interaction in Spin Triplet States

The spin-spin interaction in most ions is small compared to the spin-orbit interaction, but it becomes important when the spin-orbit interaction is missing. It arises from the magnetic dipole-dipole interaction between the spins and, expressed in terms of electron spin operators $\mathbf{s}_n$ and their relative positions $\mathbf{r}_{nn'}$, it takes the form of a double sum over the electrons

$$\mathcal{H}_{SS} = 4\mu_B^2 \sum_{n<n'} \left[\frac{\mathbf{s}_n \cdot \mathbf{s}_{n'}}{r_{nn'}^3} - \frac{3(\mathbf{r}_{nn'} \cdot \mathbf{s}_n)(\mathbf{r}_{nn'} \cdot \mathbf{s}_{n'})}{r_{nn'}^5}\right]. \qquad (6.63)$$

For triplet states there are only two electrons, so the summation can be left out and (n, n') replaced by $(1, 2)$. This can be rewritten in tensor notation, according to Judd (1963) p. 88, as

$$\mathcal{H}_{SS} = -\frac{12\sqrt{5}\mu_B^2}{r_{12}^5}\{\{\mathbf{r}_{12}\mathbf{r}_{12}\}^{(2)}\{\mathbf{s}_1\mathbf{s}_2\}^{(2)}\}^{(0)}. \tag{6.64}$$

The meaning of this formula is that the $\mathbf{r}$ and $\mathbf{s}$ operators are each to be coupled together to give a second-order spherical tensor, and then the scalar product of these tensors is to be taken. A second-order tensor goes over unsplit into an H representation of the icosahedral group, so this formula can be used effectively unaltered. As one of these tensors operates only on the space coordinates and the other only on the spins, they can be treated separately. The spin part acts within the spin triplet state as the H part of the operator $(S_i S_j + S_j S_i)$; the space part will also operate as an H-type operator within the orbital states; and the whole will be multiplied by a reduced matrix element that will depend on the details of the charge distribution in the electronic orbitals. The application of the spin-spin Hamiltonian to the various systems then follows exactly the same path as the $(\mathbf{L} \cdot \mathbf{S})^2$ operator described above. If the ground electronic state is nondegenerate T_2, then the spin-spin interaction produces D and E terms in the spin Hamiltonian, though at the Jahn-Teller minima the symmetry is such that $E = 0$. The principal values of this D-tensor are proportional to $(\mathcal{E}_1, \mathcal{E}_2, \mathcal{E}_3)$, the roots of the equation (6.58). The directions of the principal axes can be predicted by symmetry when the distortion is D_{5d}, D_{3d} or D_{2h}, but otherwise must be found from the eigenvectors of the matrix. If the ground electronic state is a nondegenerate T_1, then $E = 0$ always and D is constant, so the apparent symmetry is always axial. As the space tensor is an H-type operator within the Jahn-Teller ground state, the spin-spin energy term will always be quenched by $K(H)$. This feature is in contrast to the quadratic L_{ij} term, which arises from off-diagonal matrix elements. If the ground state is T_1 or T_2 with strains sufficiently small so that the threefold degeneracy persists into the vibronic ground state, then the spectrum will be as described for spin-orbit coupling, except that the g factor is unaffected. If g is isotropic, the spectrum for $^3T_2$ will consist of three pairs of lines that are on either side of the $g = 2$ position at distances proportional to the roots of the equation (6.58), $\mathcal{E}_1, \mathcal{E}_2, \mathcal{E}_3$, with (ℓ_x, ℓ_y, ℓ_z) being the direction cosines of the field. The spectrum for $^3T_1$ will be similar, but isotropic with two of the sets of lines coinciding.

6.2.5 Esr on C_{60}: Experiment and Theory

Many spin resonance experiments have been done on C_{60} in various surroundings and charge states, and the results will be outlined below. This work is still in a very preliminary state, and authors disagree about the origins of the spectra and about what model should be used in their interpretation. In this account we are in no position to give a detailed discussion of, or draw conclusions from, this body of work. Therefore we shall cite few references and reserve a fuller list to the bibliography. Neutral C_{60} has a closed shell configuration, so for esr to be seen in a ground state, the molecule must carry one or more extra electrons. As shown in Figure 1.3, the lowest empty orbital is of t_{1u} symmetry, so the ground configuration of $C_{60}{}^{n-}$ before the Jahn-Teller is introduced should be t_{1u}^n. In the neutral molecule, spin resonance can be seen in a metastable excited state with a spin of 1.

6.2.5.1. $C_{60}{}^-$

This ion, with one t_{1u} electron outside closed shells, is in a $^2T_{1u}$ state. A calculation of the energy levels in the free molecule has been done by Koga and Morokuma (1992) in a way that permits relaxation of the shape of the cage to either D_{5d}, D_{3d}, or D_{2h} symmetry. They find that the lowest energy corresponds to a D_{5d} distortion, extended along the symmetry axis; D_{2h} comes next, and then D_{3d}. All these energies are about 2 kcal/mole below the undistorted energy, and the difference between them is about ten times smaller than that. This corresponds to a model for the system $T_{1u} \otimes h$ with a warping potential that is small compared with the Jahn-Teller energy, with a sign that produces icosahedral minima. The calculation predicts distortions extended along the D_{5d} axes. An account of the esr spectra in various surroundings is given by Stinchcombe et al. (1993). The measured g factors are usually a little below the spin g factor, though the differences are so small that it is necessary to remember that the spin g factor is 2.0023, not 2.0000. At low temperatures, structure is seen that is interpreted as corresponding to a difference between $g_\parallel$ and $g_\perp$ in the spin Hamiltonian (6.47); in one compound $g_\parallel = 2.0023$ and $g_\perp = 1.9968$. This observation would fit with the system being trapped by strain or crystal fields in a set of minima. It would also fit with the g factors for the pseudo-rotational state, (6.55). As these are powder spectra, averaged over direction, the two possibilities cannot be clearly distinguished, though as the molecular surroundings are almost bound to be of lower than icosahedral symmetry, the pseudo-rotational model is unlikely to hold good. The structure averages out at a temperature of about 50 K.

6.2.5.2. $C_{60}{}^{2-}$

The theory described in Chapter 3 Section 3.5 shows that the configuration t_{1u}^2 may have either a spin singlet or triplet ground state. Furthermore the calculations made by Negri, Orlandi, and Zerbetto (1992) find a considerable amount of the next excited orbital, t_{1g}, mixed in by configuration interaction, with a concurrent reordering and compression of the term energies even before the Jahn-Teller interaction is included (see Section 6.3.2). Experimentally it is hard to decide whether the ground state is singlet or triplet, and both conclusions have been reached by different groups. For two different conclusions and details of the experimental techniques and difficulties, see papers by Boyd et al. (1995)

and Trulove et al. (1995). The singlet and triplet states are evidently rather close in energy. The spectra typically show a superposition of triplet spectra, corresponding to the spin Hamiltonian (6.46), averaged over direction. The different triplet spectra have small but differing values of the parameter D. The g factors are below the free spin value but higher than those for $C_{60}{}^{-}$. In addition there are sharp lines with no terms in D and E, and it is not certain whether they belong to the same species. In one compound whose structure has been determined by X-ray analysis (Paul et al., 1994), the $C_{60}{}^{2-}$ ion is found to be extended along a D_{5d} axis to a good approximation, but the exact symmetry is only C_i. With surroundings of such low symmetry, it is doubtful how much the distortion owes to the Jahn-Teller interaction, though the magnitude of the distortion is similar to that calculated by Koga and Morokuma (1992) for free $C_{60}{}^{-}$. This may be a case where the Jahn-Teller effect dictates the size and type of the distortion, and the surrounding crystal field picks out the direction and causes some small extra distortion as well.

6.2.5.3. $C_{60}{}^{3-}$

For this spectrum see a paper by Khaled at al. (1994). It shows structure, which is wiped out at about 50 K. The average g factors are very nearly free-spinlike. There are also sharp lines at $g = 2$ that may belong to other species, and the complex is relatively unstable. The spectrum arising from the ^{4}A state should have an isotropic g factor, but configuration interaction could produce nonzero D and E parameters in the spin Hamiltonian if the symmetry is reduced. It is worth noting that the theory of Chapter 3 predicts a rotating orthorhombic distortion for the ground state of the Jahn-Teller coupled spin doublet terms, unlike the rotating axial distortion predicted for all the other charge states. If such a distortion were frozen in by crystal fields, a spin Hamiltonian for the doublet states with three different g factors might result (6.48).

6.2.5.4. $^{3}C_{60}$

The lowest excited state of C_{60} is a spin triplet, and spin resonance is observed in it. The state is formed by optical excitation to a spin singlet state, which then decays into the spin triplet state. Theory suggests that both these states are of T_{2g} type. The lowest energy excitation takes an electron out of a filled h_u shell into a t_{1u} orbital. Coupling the hole and electron can produce the terms $^{1,3}T_{1g}$, $^{1,3}T_{2g}$, $^{1,3}G_g$, and $^{1,3}H_g$. Calculations by Negri, Orlandi, and Zerbetto (1988) put the triplets below the singlets, as would be expected, and find that T_{2g} is the lowest state in each series. The other predicted states come fairly close, and odd symmetry terms arising from promotion to the t_{1g} orbital appear among them. The Jahn-Teller interaction in this case has been studied numerically by Surján, Udvardi, and Németh (1994). These authors have calculated the energies with D_{5d}, D_{3d}, and D_{2h} distortions, and found the lowest energies to be 185 meV, 59 meV, and 120 meV respectively below the energy of the undistorted complex. This would represent a weak Jahn-Teller interaction with strong warping in a $T_2 \otimes h$ system, with the minima at D_{5d}. One way in which this sign of the warping might arise is through a Jahn-Teller coupling between states of different energy, which is sometimes called a pseudo-Jahn-Teller effect. The effect of this on the T_2 states can be introduced as a second-order perturbation that will have the term splitting as a denominator. Within the T_2 states, this can be written as a linear combination of two operators, $h_A^{(2)}$ and $h_H^{(2)}$, which are quadratic in the h vibrational coordinates and are chosen to be overall operators of A and H symmetry. If this perturbation is applied to the ground state of a $T_2 \otimes h$ system, it produces warping of the lowest APES with the usual angular distribution, V_{icos}. The minima are at D_{5h} distortions, whatever the relative signs of the Jahn-Teller interactions. If the Jahn-Teller coupling within T_2 is small compared with the pseudo-Jahn-Teller coupling, and the splitting between the terms is small, then this situation, that of the warping energy being comparable to the Jahn-Teller energy, could happen. Experiments have been done using various techniques: esr (Wasielski et al., 1991), optically detected magnetic resonance, light-induced esr (Lane et al., 1992) and electron spin echo (Groenen et al., 1992). Most of the experiments have been done in disordered surroundings; there is agreement in fitting the spectrum to a triplet spin Hamiltonian (6.46) with an isotropic $g = 2$ and $|D| = 0.0114$ cm^{-1}, $|E| = 0.00069$ cm^{-1}. The conflict with the theory here is immediately apparent. All the models so far proposed for T states have D_{5d} minima, and systems trapped at these points would have $E = 0$ axial symmetry. A calculation by Surján et al. (1996) finds $D = -0.009$ cm^{-1}, $E = 0$ at the D_{5d} points. There is an additional complication in the interpretation of the esr spectra in excited states compared with the ground states, and it comes about because the spin system does not have time to achieve thermal equilibrium in the course of the experiment. The spectrum observed will always depend on the relative occupation levels of the various spin states, which in the ground states corresponds to thermal equilibrium. In an excited state the populations will depend on the pathway by which the state is reached, and it is assumed that no further equilibration takes place before the esr measurement is made. This will result in population inversions leading to some stimulated emission lines as well as absorption lines in the spectrum, which can considerably enhance the visibility of the spectrum compared with ground state esr. The spectrum of $^3C_{60}$ can be simulated by using the values of $|D|$ and $|E|$ already quoted, with the assumption that the only populated state is $S_z = 0$, where the z direction is the z direction in the spin Hamiltonian. Simulation is necessary because the axes may take any orientation with respect to the field. An account of a simulation process for $^3C_{60}$ is given by Regev et al. (1993) in which they include the effect of a relaxation process that involves a reorientation of the spin Hamiltonian axes and that increases with temperature. Experimentally the structure disappears somewhere between 38 K and 64 K. The physical mechanism for producing the particular choice of initial spin populations that gives the right band shape remains unexplained. The spin populations appear to be aligned in the course of the singlet to triplet transition with respect to axes chosen by the Jahn-Teller distortion. This, on the face of it, suggests that the $^3T_{2g}$ distortion is already defined before the transition, so that it has something to do with the Jahn-Teller distortion in the even more short-lived $^1T_{2g}$ state. This spectrum has been attributed to the system's being trapped in a set of Jahn-Teller minima of D_{2h} symmetry, but none of the foregoing theory predicts the existence of such minima. Another difficulty in this model can be seen in Figure 6.2; the eigenvalues plotted there should also be the principal values of the D tensor, but the eigenvalues at the D_{2h} point imply a considerably larger value of $|E/D|$ than the measured one. It is very difficult to think of a set of parameters that would produce such minima or the measured value of $|E/D|$ at these points, and none have been proposed as yet. It may be that the answer will lie in the pseudo-Jahn-Teller effect and the closeness of orbitals of different symmetries. To direct this discussion it is worthwhile looking at Figure 6.3, which plots the predicted value of $|E/D|$ over the spherical trough produced by the $T_2 \otimes h$ Jahn-Teller interaction. The quantity plotted is found by inspecting the eigenvalues of the matrix (6.57), taking the two eigenvalues that are closest to each other to be equal to $\frac{1}{3}D \pm E$, and the other eigenvalue to be $-\frac{2}{3}D$. The first contour around the D_{5d} and D_{3d} zeros corresponds to the experimental value of $|E/D|$, and the maximum value of $|E/D|$, which appears white on the plot, represents the boundary on which D changes sign. It is fairly clear that an average over all directions, such as might be produced by random strains, will be featureless because positive and negative D are equally likely. Crystal fields that select those particular points on the APES with the measured value of $|E/D|$ would produce the spectrum, but it is hard to see how such specific fields could arise in a random matrix. One way of shifting the wells on a Jahn-Teller APES is to introduce terms that are of higher order in the vibrational coordinates. The warping has already been introduced as the effect of a term that is cubic in the coordinates, which produces a sixth-order harmonic function, V_{icos}, on the spherical APES (Section 3.2.3). This function is also the A_g component of $L = 5$. Inspection of Table 2.4 shows that the next A_g harmonic corresponds to $L = 10$, or five powers of the coordinates. There do exist fourth-order icosahedral invariants of the h coordinates, but they do not introduce a different shape to the warping. This is one reason why it is so unlikely that perturbations will make a radical change to the positions of the wells unless the electronic orbitals are so close to each other that perturbation theory is invalid. Just the same, in the absence of better calculations, it is worth looking at the effect of higher order terms, and one possible result is shown in Figure 6.4. Here the effect of adding a small amount of V_{icos}^2 to $-V_{icos}$ is plotted. The amount was chosen so as to replace the single D_{5d} minimum of

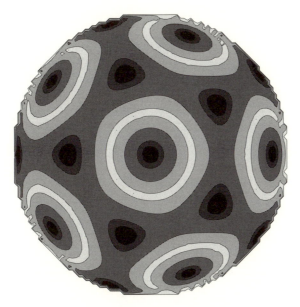

Figure 6.3. A contour plot of the ratio $|E/D|$ calculated from the eigenvalues of the matrix (6.57) on the spherical minimum. The zeros at D_{5d} and D_{3d} appear black, and the first contour around them corresponds to the experimental value of $|E/D|$. The maximum value of $|E/D|$ appears white on the plot and represents the boundary on which D changes sign. Contours are at $|E/D| = (69/1140, 0.1, 0.2, 0.3)$

$-V_{icos}$ by a (nearly) circular trough. A section through that trough is shown in Figure 6.5. While this picture does not represent a sensible perturbation scheme, and V_{icos}^2 contains other than the $L = 10$ harmonic, it does indicate the sort of minimum that could produce the observed result in conjunction with random strains or crystal fields. The strains would then act to pick out different points on the minimum for different molecules, and all these points would have approximately the same value of $|E/D|$. In the experiments on single crystals of C_{60} done by Groenen et al. (1992), the values of $|D|$ and $|E|$ can be found directly rather than being unfolded from a band shape. Two distinct spectra were found with their principal axes related to the cubic symmetry axes of the crystalline structure, and with different values of $|D|$ and $|E|$. Their results can be set out as follows:

Axes	Triplet α	Triplet β	C_{60} in Glass
[001]	140 ± 1	39 ± 1	228
[110]	8 ± 1	171 ± 1	93
[1$\bar{1}$0]	132 ± 1	132 ± 1	135
$|E/D|$	0.295	0.181	0.063

(6.65)

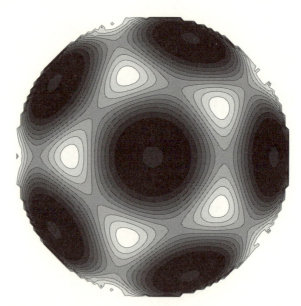

Figure 6.4. A contour plot of $-V_{icos} + V_{icos}^2/20$ over the spherical Jahn-Teller surface, showing a trough surrounding the D_{5d} point. The contour shading makes minima black and maxima white.

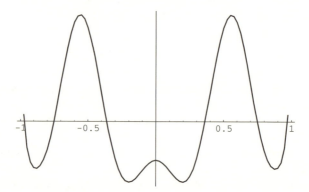

Figure 6.5. A plot of the energy on a section through the center of Figure 6.4, showing the profile of the energy trough around the D_{5d} point at the center. The section is taken along the vertical axis of that figure.

Here the directions of the principal axes of the D tensor are listed, along with the principal values along those directions. The principal values found in the experiments on C_{60} in a random matrix are given alongside, expressed in the same units. The overall magnitudes are similar, but the different values of $|E/D|$

direct us to different places on the sphere in Figure 6.3. The explanation favored by the authors (Groenen et al.) is the spread of the excitation over pairs of neighboring molecules. Another possibility is that the positions on the spherical surface are being determined by crystal fields related to the surrounding crystal structure. Later work has found some extra triplet states in the crystal, some of which may be attributable to defects.

6.3 ENERGY LEVELS IN C_{60}^{n-}

The theory of the systems $p^n \otimes h$ that was given in Chapter 3, Section 3.5, is particularly important in connection with the negatively charged states of C_{60}^{n-}. Two questions arise as extra electrons are added to the C_{60} molecule. First, what is the nature of the ground state, that is, are the electrons arranged in a high or low spin configuration? Second, is there a negative or positive change in the energy as more electrons are added? Section 3.5 gives the answer to both of these questions in principle, but requires discussion of how it should be applied.

6.3.1 C_{60} Vibrational Modes and Their Coupling

The sixty atoms comprising C_{60} give rise to $(180 - 6)$ modes of vibration, so there are necessarily multiple modes of every symmetry type. The full list is

$$2a_g \oplus 3t_{1g} \oplus 4t_{2g} \oplus 6g_g \oplus 8h_g \oplus a_u \oplus 4t_{1u} \oplus 5t_{2u} \oplus 6g_u \oplus 7h_u. \quad (6.66)$$

The large task of calculating the energies and normal coordinates of these modes has been undertaken by several groups of workers, but perhaps the most enlightening way of organizing the information is that proposed by Ceulemans et al. in 1993. These authors pointed out that the calculated modes could be quite well approximated by the vibrations of a hollow sphere, and in particular that the lowest mode, of h_g symmetry, is the simple "squashing" deformation derived from angular momentum states with $L = 2$. They remark that this mode has the same symmetry as tidal waves on a flooded planet. It is also related to the form of h_g mode used in the pictures of distortions in Figures 2.5 and 2.7–2.9. The Jahn-Teller couplings of these modes with an electron in a t_{1u} orbital have also been calculated by various groups, and we show here in Table 6.2 some of the results, extracted from Table II of Manini et al. (1994). Two key facts emerge from this table: first, there is an extensive spread of frequencies among the h_g modes, with a factor of about 6 between the highest and the lowest; second, even though the calculations differ in detail, they agree in spreading the Jahn-Teller interaction strength rather evenly across the whole range of modes. These are the circumstances in which the cluster model developed in Section 6.1 is needed. The best approximation to the ground states is then

TABLE 6.2

Coupling Strengths for h_g Modes from Manini et al. (1994)

Mode energies are experimental, from Zhou et al. (1992), and the coupling constants are from Antropov et al. (1993) (k'), Varma et al. (1991) (k'') and Schlüter et al. (1992) (k''').

Mode	Energy in cm^{-1}	k'_n	k''_n	k'''_n
1	270.0	0.33	0.33	0.54
2	430.5	0.37	0.15	0.40
3	708.5	0.20	0.12	0.23
4	772.5	0.19	0.00	0.30
5	1099.0	0.16	0.23	0.09
6	1248.0	0.25	0.00	0.15
7	1426.0	0.37	0.48	0.30
8	1575.0	0.37	0.26	0.24
$\sqrt{\sum k_n^2} =$		0.83	0.70	0.88
$\sum k_n^2 \omega_n =$		661.5	542.5	510.5

given by $k_{\text{eff}} = \sqrt{\sum k_n^2}$ and $k_{\text{eff}}^2 \omega_{\text{eff}} = \sum k_n^2 \omega_n$. These quantities are included in Table 6.2. Reference to the numerical calculations in Figure 6.1 shows that if k_{eff}^2 is a little less than one, the Jahn-Teller energy is a little less than $\hbar \omega_{\text{eff}}$. The Jahn-Teller energy calculated by Koga and Morokuma (1992) is about 690 cm^{-1}, which is a surprising match, considering all the uncertainties.

6.3.2 Configuration Interaction in C_{60}^{n-}

Another respect in which C_{60} differs from the model in Section 3.5 is that the t_{1u} orbital is not alone. The next unoccupied orbital above t_{1u} is t_{1g}, and it is rather close in energy. The result is that configurations other than p^n must be allowed for. The effect of this on the term energies has been investigated by Negri, Orlandi, and Zerbetto (1992). These authors have done calculations of the energies of the low excited states of C_{60}^{n-} up to $n = 6$ with the inclusion of the t_{1g} states. They find a different ordering of terms from the Hund's rule ordering of Section 3.5.4, as well as extra terms deriving from other configurations, particularly when $n > 3$ and higher spin states appear. They point out that the energies are all so close that other types of calculation may well give other orderings, but it is worth listing a few of their results in Table 6.3 to show the sort of thing that happens: here the energies are in eV, and the terms that would be expected from applying Hund's rule to p^n are shown in parentheses. The amount of configuration interaction indicated by the energies in this table will alter the

TABLE 6.3

Calculated Term Energies for the Configuration p^n

Term energies (in eV) for the configuration p^n calculated by Negri, Orlandi, and Zerbetto (1992), showing the effect of configuration interaction

n					
2	^1A_g	^1A_u	$^3T_{1g}$	^1H_g	
	0	0.16	0.19	0.30	
	(^1S)		(^3P)	(^1D)	
3	^4A_u	^4A_g	$^2T_{1u}$	^2H_u	^4H_g
	0	0.10	0.42	0.43	0.45
	(^4S)		(^2P)	(^2D)	
4	^1A_g	$^3T_{1g}$	^1A_u	^1H_g	$^1T_{1u}$
	0	0.01	0.14	0.15	0.16
	(^1S)	(^3P)		(^1D)	
5	^4A_g	$^2T_{1u}$	$^6T_{1u}$	^4A_u	^4T_g	^4H_g
	0	0.01	0.11	0.19	0.21	0.24
		(^2P)				

model used in the present calculations in several ways. The accidental SO(3) symmetry will disappear and with it the simple way all the coupling coefficients depend on one parameter, k. The Jahn-Teller coupling coefficients within and between terms will additionally depend on the coupling within t_{2g}, and there will be pseudo-Jahn-Teller coupling via odd modes of vibration between odd and even terms with the same spin. It is possible to make some general remarks about the effect of the Jahn-Teller interaction on the terms shown in Table 6.3. The magnitude of the Jahn-Teller energy for C_{60}^- is about 0.086 eV (Koga and Morokuma, 1992). The results of Section 3.5 show that the Jahn-Teller energy should be four times as large as in C_{60}^- in the spin singlet states of p^2 and p^4, and three times as large in the spin doublet states of p^3. Thus the Jahn-Teller energies and the term splittings are comparable in size.

In $n = 2$ and $n = 4$, the configuration interaction already makes the $^1A_g(^1S)$ state lowest, contrary to the Hund's rule prediction, and the Jahn-Teller interaction should enhance that energy difference, following the calculations plotted in Figure 3.7. In $n = 4$ the near degeneracy of $^1A_g(^1S)$ and $^3T_{1g}(^3P)$ will allow a linear Jahn-Teller coupling within the triplet state to make it the ground state if that coupling is really weak, but at the intermediate coupling strength assumed, the pseudo-Jahn-Teller coupling between $^1A_g(^1S)$ and $^1H_g(^1D)$ can be expected to bring $^1A_g(^1S)$ to the bottom. In $n = 3$ the Hund's rule that puts a quartet state lowest is obeyed. There is the possibility of a pseudo-Jahn-Teller coupling between ^4A_g and ^4H_g states to push this ^4A_g state down and compete with the level crossing shown in Figure 3.9. In $n = 5$ the number of nearby quartet states could make the effect of off-diagonal coupling of the ^4A_g state competitive with the effect of linear coupling of $^2T_{1u}(^2P)$. We are able to report some experimental evidence for the ordering of the states in the section on spin resonance, Section 6.2.5.2.

6.4 MOLECULAR SPECTRA

The influence of the Jahn-Teller effect on molecular spectra derives from its ability to lift degeneracies in electronic energy levels, thereby altering the distribution and energies of spectral lines. Line spectra from Jahn-Teller complexes typically display irregular energy level spacings quite different from the uniform spacings obtained when molecular vibrations and electronic states are not coupled in this way. Though distinct in appearance, the spectral evidence of the Jahn-Teller effect can be hard to pin down due to competing interactions, and the many degrees of freedom typical of icosahedral molecules or defects can only complicate the task. Given the complexities, we think it important to emphasize several facts central to how spectra are affected by Jahn-Teller interactions: (i) symmetry-lowering distortions (such as occur in the "static" Jahn-Teller effect) typically will not be stable unless locked in place by an outside perturbation; (ii) there can be no transitions between states split in energy by a Jahn-Teller interaction except those assisted by an odd-symmetry (*ungerade*) phonon. We will discuss each of these points in turn.

6.4.1 Symmetry-Lowering Distortions

To say that symmetry-lowering distortions may not be stable is not to say that they will not occur, only that such distortions might not be observable, even at low temperatures. The $T \otimes h$ system provides an easy example: the minimum of the lowest APES possesses SO(3) symmetry, which gives rise to pseudo-rotational ground states. Each point on this surface (Q_0, θ, ϕ) corresponds to an axially symmetric distortion of the icosahedral complex, with the axis of the distortion specified by (θ, ϕ). In the absence of any additional symmetry-lowering interaction, all states (θ, ϕ) are equally probable, and the complex will be subject to a rotating distortion. The symmetry of this distortion will vary over the allowed symmetries of the complex, that is, the subgroup symmetries of I_h, and the distortion will average out in any observation.[2] The symmetry of the lowest APESs energy minimum will be reduced from SO(3) to I_h with the

---

[2]Though we represent the distortion in the form of an ellipsoidal surface, as in Figure 3.1, it is good to remember that an icosahedral complex will have atoms and bonds overlaying this ellipsoid. It is these physical features that distinguish the symmetry of the distortion, D_{5d} from D_{3d}, for example.

addition of a warping potential such as $V_{icos}{}^3$; however, this will not lower the configuration symmetry of the molecule unless the warping-induced minima are deep enough to quench inter-well tunneling. As Section 3.2.3 shows, tunneling will produce a superposition of localized states, either a T_1 or a T_2 state for $T \otimes h$, and the molecule will retain I_h symmetry. In order to reduce the symmetry of the molecular complex, a lower-symmetry perturbation is needed, such as might arise from the local environment of the molecule (e.g., crystal fields or strains).. Even a small lower-symmetry term can lower the energy of one minimum sufficiently so as to trap the wave function in a single well. Whether this is a D_{5d} or a D_{3d} minimum will depend on the sign of V_{icos}, but the important point is this: no "static" Jahn-Teller distortion can occur without an external symmetry-lowering distortion to stabilize it.

6.4.2 Allowed Transitions

The Jahn-Teller interaction has the symmetry of the molecular configuration and thus cannot remove a degeneracy in energy. It can lift an electronic degeneracy, but it can only replace an electronic degeneracy with a vibronic degeneracy—an electronic triplet is superseded by a vibronic triplet, as we saw in Section 6.4.1. The symmetry of the interaction also means that it will split even-symmetry states (*gerade*) into states of the same symmetry; likewise for odd-symmetry states (*ungerade*). The effect of this preservation of inversion symmetry is as given at the beginning of Section 6.4: there will be no transitions between states that arise from an electronic degeneracy lifted by the Jahn-Teller effect, except for transitions mediated by an odd-symmetry phonon. This restriction follows from the selection rules for electric dipole transitions.

6.4.2.1. Selection Rules

Transitions between energy states in molecules typically are termed *first-order* if they are mediated by the electric-dipole operator but involve no vibrational excitations. Molecular vibrations are created or destroyed in what are called *vibronic transitions*—second-order processes that often are the dominant absorption mechanism in molecules. Group theoretical arguments determine the selection rules for both processes. Transitions involving the emission or absorption of electric dipole radiation require nonzero values for the matrix element of the electric dipole operator, $\mathbf{r} \cdot \mathbf{E}$, between the initial and final molecular states. Labeling states by irreps of the molecule's symmetry group, this matrix element has the form

$$\langle \Psi(\Gamma_i) | \mathbf{r} \cdot \mathbf{E} | \Psi(\Gamma_f) \rangle, \tag{6.67}$$

---

[3] See Equation (3.14).

where i and f indicate *initial* and *final*. If the molecule has icosahedral symmetry, then we should note that $\mathbf{r}$ transforms as the T_{1u} irrep in this symmetry. This means that the selection rule for an allowed zero-phonon transition reduces by symmetry arguments to the requirement that T_{1u} be contained in the direct product $\Gamma_i \otimes \Gamma_f$. For vibronic transitions, we must take account of the symmetry of vibrations involved in the transition. If there is a single quantum of vibrational energy involved in the transition, then symmetry requirements are met by the following selection rule: if the vibrational mode is labeled by irrep Γ_v, then the direct product $\Gamma_i \otimes \Gamma_v \otimes \Gamma_f$ must contain irrep T_{1u}. If there is a "static" Jahn-Teller effect, such that the icosahedral molecule is locked into some lower symmetry distortion, then the irreps in the above selection rules will be those for the group of lower symmetry. As Figure 2.4 shows, a single g or h vibrational mode can lower the symmetry from that of I_h to that of any of the following I_h subgroups: T_h, D_{2h}, D_{3d}, or D_{5d}; this distortion may be stabilized by an external influence. In these lower symmetry environments, we must use the appropriate irrep for $\mathbf{r}$, but otherwise the group theoretical conditions for transitions will have the same form as for I_h. The T_{1u} irrep for I_h transforms under these lower symmetries as follows:

$$I_h \ : \ T_{1u}$$

$$
\begin{aligned}
D_{5d} \ &: \ A_{2u} \oplus E_{1u} \\
D_{3d} \ &: \ A_{2u} \oplus E_u \\
D_{2h} \ &: \ B_{1u} \oplus B_{2u} \oplus B_{3u} \\
T_h \ &: \ T_u.
\end{aligned}
\tag{6.68}
$$

Thus for a D_{5d} distorted complex, selection rules will allow a zero-phonon transition whenever $\Gamma_i \otimes \Gamma_f$ contains either an A_{2u} or an E_{1u} irrep (all irreps for D_{5d}). Similarly, single-phonon vibronic transitions between states Γ_i and Γ_f will be allowed if A_{2u} or E_{1u} is contained in the decomposition of $\Gamma_i \otimes \Gamma_v \otimes \Gamma_f$. Selection rules for the other symmetries follow easily from (6.68).

6.4.3 Experimental Evidence

The Jahn-Teller effect is invoked often as a contributing factor in the spectra of icosahedral complexes, but clear experimental evidence of the exact form of its involvement can be obscured by the spectral complexities that naturally arise in systems with many degrees of freedom. The strong interest in C_{60} has led to a good deal of experimental investigation of the spectra of its anions and cations. The anions C_{60}^{n-} ($n = 1, \ldots, 5$) and cations C_{60}^{n+} ($n = 1, \ldots, 9$) should each have a disposition towards a Jahn-Teller distortion. The singly-ionized cation and the anions have recieved the most study, and of these C_{60}^{-} and C_{60}^{+} have the clearest and best-defined spectra. In what follows, we will review the spectral evidence for Jahn-Teller interactions in these two ions. Experimental work on C_{60} and its ions is being vigorously pursued at present, and clearly there is much more information to come.

6.5 C_{60}^- SPECTRA

The spectrum of the C_{60}^- anion has been extensively investigated and offers a detailed field of data to consider with regard to Jahn-Teller activity. This anion has a distinct spectral signature in the near-infrared (near-IR) region, in contrast to C_{60}, which is featureless in this region. The T_{1u} LUMO of C_{60} is singly occupied for C_{60}^-,[4] prompting a $T \otimes h$ Jahn-Teller instability as discussed in Chapter 3. Constraints on experimental designs usually result in the anion being placed in solutions (e.g., in CH_2Cl_2, Heath et al., 1992) or as defects in crystal lattices (e.g., in argon, Gasyna et al., 1992). Such environments typically produce small shifts in the absorption spectrum of C_{60}^- but do not appear to alter the band structure otherwise; we should expect Jahn-Teller effects to be robust in such environments if they are favored energetically. An example of the effects of different environments and temperatures is seen in Figure 6.6, which shows the electronic absorption spectrum for C_{60}^- in the near-IR region. The correspondence of the principal bands, whether from within a neon, argon, or MCH (methylcyclohexane) matrix, is easily seen (Kondo et al., 1995). Above about $10300\ \mathrm{cm}^{-1}$, the spectrum difference between these three spectra and that produced from the MTHF (2-methyltetrahydrofuran) matrix has been interpreted by Kondo et al. (1995) as evidence for environmentally stabilized D_{2h} distortions of the C_{60}^- anion. They observe that the MTHF matrix differs from the other three in that it is polar, and a consideration of the allowed transitions in each symmetry has led them to suggest that an additional perturbation in the MTHF matrix acts to stabilize the anion in a distortion different from those obtained for the nonpolar argon, neon, and MCH matrices. Kondo et al. (1995) conclude that C_{60}^- distorts to D_{2h} symmetry in polar media such as MTHF and to D_{5d} and/or D_{3d} symmetry in nonpolar media such as rare gas matrices. Implicit in their analysis is the assumption that C_{60}^- is distorted to D_{5d} or D_{3d} when embedded in a rare gas or MCH matrix. As we have seen, both symmetries do arise as energy minima for icosahedral warping, but there needs to be an additional lower symmetry perturbation if these distortions are to be stable. The rare gas and MCH environments thus must exert a symmetry-lowering perturbation on C_{60}^-—even though they are nonpolar—if these distortion symmetries are to be realized. The spectra are consistent with a D_{5d} or D_{3d} symmetry, and it is reasonable to suppose that the stability of these distortions arises from some additonal symmetry-lowering

---

[4]See the single-electron energy levels in Figure 1.3.

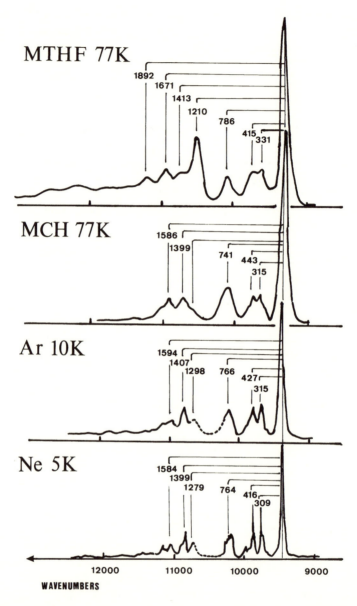

Figure 6.6. Absorption spectra for C_{60}^{-} in the near-IR region. Labels indicate the matrix in which the anion was placed and the temperature at which the spectrum was taken (MTHF = 2-methyltetrahydrofuran and MCH = methylcyclohexane). The neon and argon data are taken from Fulara et al. (1993), and the dotted curves indicate the region where a superposed absorption spectrum for C_{60}^{+} has been subtracted away. Numbers labeling bands indicate the shifts from the 0-0 band in cm^{-1} (Figure 1 from Kondo et al., 1995). Spectra have been shifted (slightly) so that the 0-0 peaks are aligned.

perturbation within the rare gas and MCH matrices. The data in Figure 6.6 for argon and neon matrices comes from Fulara et al. (1993), who identify the near-infrared spectra as due to an $E_{1g} \leftarrow A_{2u}$ electronic transition in D_{5d} symmetry. This symmetry is in keeping with the ab initio calculations of Koga and Morokuma (1992) and indicates a Jahn-Teller distortion in which the T_{1u} ground state is split to E_{1u} and A_{2u} states, with A_{2u} the new ground state (see Section 6.2.5.1). The T_{2g} excited state splits under this symmetry such that $E_{1g} \leq A_{2g}$, and this paves the way for a zero-phonon transition (since E_{1u} is contained in $A_{2u} \otimes E_{1g}$). The higher-energy transitions (i.e., those lines shifted by 309, 416, and 766 cm^{-1} in neon) can be understood in terms of one-phonon vibronic transitions: the h_g vibrational modes in the I_h symmetry of C_{60} each yield a totally symmetric component (A_{1g}) in D_{5d}, and the observed absorption spectrum for the neon matrix has a close one-to-one correspondence with the h_g mode frequencies (see Table 6.4). The D_{2h} distortion symmetry proposed by Kondo et al. for the anion in a MTHF matrix must also arise as a result of an additional perturbation. (Icosahedral warping does not produce energy minima with this symmetry.) Stinchcombe et al. (1993) have pointed out that D_{2h} symmetry could be induced by an interaction at one of the olefinic double bonds of C_{60}^- (i.e., a bond between two hexagons). If the D_{2h} distortion symmetry is stable, it must be due to this or a similar perturbation that arises in the polar MTHF matrix.

6.6 C_{60}^+ SPECTRA

In icosahedral symmetry, the singly-charged cation C_{60}^+ possesses an ^2H_u ground state and ^2H_g and ^2G_g low-lying excited states. Fulara et al. (1993) have measured the absorption peaks in the spectrum for C_{60}^+ isolated in neon matrices at 5 K, and their data is given in Table 6.5. Similar absorption spectra for C_{60}^+ in argon and in freon have been reported (Gasyna et al., 1992; Kato et al., 1991) with equivalent progressions but for small matrix-related shifts. The ^2H_u ground state and the ^2H_g and ^2G_g excited states are all capable of undergoing Jahn-Teller distortions along h_g and g_g normal coordinates, leading to D_{5d}, D_{3d}, or D_{2h} distortions, as outlined in Chapters 4 and 5. INDO[5] calculations by Bendale et al. (1992) on C_{60}^+ predict a $^2A_{1u}$ ground state in D_{5d} symmetry stabilized 8.1 kcal/mol below the energy of the icosahedral configuration, with 2.2 kcal/mol pseudo-rotation barriers at points of D_{2h} symmetry. This pattern of minima and saddle points arises in the $H \otimes (g \oplus h)$ system discussed in Section 5.8, with $E^1 > 0$. In D_{5d}, the H_u level splits into three levels: $A_{1u}, E_{1u},$ and E_{2u}; the H_g follows the same pattern to form *gerade* levels. Bendale et al. (1992) have calculated the $^2E_{1g} \leftarrow {}^2A_{1u}$ transition at 8630 cm^{-1},

[5]Intermediate Neglect of Differential Overlap (Head and Zerner, 1985).

TABLE 6.4

Absorption Spectra for C_{60}^{-} in Neon and Argon Matrices

The assignments are to the a_{1g} modes in D_{5d} symmetry, and weak peaks are marked with a w. The Ne peak at 10859 cm^{-1} overlaps with the C_{60}^{+} band.

Ar matrix (cm^{-1})	Ne matrix (cm^{-1})	Δ (Ne matrix)	Assignment	C_{60} groundstate frequencies
9369	9943			
	9449			
	9455	0	0_0^0	
	9479			
9592	9655w			
	9671w	222		
	9678w			
9678	9726			
	9734			
	9758	309	ν_8	270(h_g)
	9764			
	9773			
9790	9850			
	9865	416	ν_7	431(h_g)
	9869			
	9872w			
	9889w			
	9930w			
	9943w			
9895	9964	514	ν_2	497(a_g)
	9971			
10129	10173			
	10213	764	ν_5	773(h_g)
	10273			
	10555	1106	ν_4	1099(h_g)
10661	10728	1279	ν_3	1248(h_g)
	10771	1322		

TABLE 6.4

continued

Ar matrix (cm^{-1})	Ne matrix (cm^{-1})	Δ (Ne matrix)	Assignment	C_{60} groundstate frequencies
10770	10848			
	10859	1399	ν_2, ν_1	$1426(h_g)$, $1469(a_g)$
	10882			
	11022			
10957	11033	1584	ν_1	$1574(h_g)$

Source: Table 2 of Fulara et al. (1993).

TABLE 6.5

Absorption Spectra for C_{60}^{+}

The observed absorption peaks in the spectra for C_{60}^{+} isolated in a neon matrix at 5 K. Wave numbers were measured as reported in Fulara et al. (1993), and the assignments were made using C_{60} mode numbering. Two progressions are shown (Δ and Δ'), the result of what appear to be two overlapping electronic transitions with origins at 10369 and 10435 cm^{-1}, possibly corresponding to two slightly different C_{60}^{+} structures within the matrix environment, one of which has D_{5d} symmetry.

Ne matrix (cm^{-1})	Δ (cm^{-1})	Δ' (cm^{-1})	Assignment
$10368^{a)}$	0		0_0^0
10435	67	0	0_0^0
10603	235		$\nu_8(h_g)$
10667		232	$\nu'_8(h_g)$
10792	424		$\nu_7(h_g)$
10845		410	$\nu'_7(h_g)$
10922		487	$\nu'_2(a_g)$
11082		647	$\nu'_6(h_g)$
11165		730	$\nu'_5(h_g)$
11715		1280	$\nu'_3(h_g)$
11841		1406	$\nu'_2(h_g)$
11955	1586		$\nu_1(h_g)$
12012		1577	$\nu'_1(h_g)$

Source: Fulara et al. (1993).

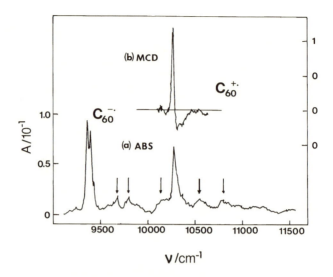

Figure 6.7. Near-IR and MCD spectra for C_{60}^{+} in an argon matrix at a temperature of 11 K. Part (a) shows the near-IR spectrum (with that for C_{60}^{-} included at left), with the MCD spectrum offset above it. Both graphs have the same horizontal energy scale. The absorption scale is at left (A = absorbance) and the MCD scale is at right with $\Delta A = A_L - A_R$ Tesla (Figure 2 of Gasyna et al., 1992). Note that energy increases to the right.

close to the observed absorption band system (whose origin is at 10368 cm^{-1} in neon). The eight h_g modes in icosahedral symmetry each have an a_{1g} component in the D_{5d} geometry, and Table 6.5 lists the likely phonon peaks associated with the C_{60}^{+} electronic band system (Fulara et al., 1993). Gasyna et al. (1992) have obtained the near-IR and magnetic circular dichroism (MCD) spectra of this cation, and both are given in Figure 6.7. The vertical arrows around the C_{60}^{+} near-IR spectrum (in the range 10300-10540 cm^{-1}) locate the vibronic sidebands for the cation system. Gasyna et al. note that the change in sign of the MCD spectrum rules out the possibility that the triplet structure could be due to cations at different matrix sites. The triplet structure and the MCD pattern for C_{60}^{+} are quite similar in form to the analogous spectra for the cases of lithium and sodium atoms embedded in xenon matrices (Rose et al., 1986), and though the use of MCD in this study adds significantly to the amount of information obtained, the theoretical interpretation of these spectra still presents a challenge.

6.7 SUPERCONDUCTIVITY IN THE FULLERIDES

Though our concern has been exclusively directed towards the Jahn-Teller effect in isolated icosahedral complexes—such as the C_{60}^{-1} anion—it is natural to wonder how such molecular interactions might be involved in the superconducting states of the C_{60}-based compounds such as K_3C_{60} and Rb_3C_{60}. To answer this question, we must consider both the C_{60} solid and the effect of doping on the C_{60} molecules in the superconducting compounds.

6.7.1 C_{60}: Molecular Crystal

Solid C_{60} possesses a face-centered-cubic (fcc) structure held together by van der Waal's binding. As such, it exists as a molecular crystal, with intermolecular vibrations generally much lower in energy than the intramolecular vibrations.[6] Even considering the large number of degrees of freedom in the unit cell, the room temperature Raman spectra for solid C_{60} is surprisingly similar to the solution spectra of the free molecule (Danieli et al., 1992). This indicates the importance of the molecular character to the physics of the solid, and we should expect to be able to model the solid starting from C_{60} as the unperturbed system.

Local-density-approximation density-functional calculations by Schlüter et al. (1992) emphasize the following for the crystal: (i) the conduction band states are derived from the T_{1u} level of C_{60}, which broadens into about a 0.5 eV-wide band; (ii) the T_{1u}-derived conduction states are principally π states, centered at the carbon atoms and pointing nearly radially outwards. The closest carbon atoms of neighboring molecules are separated by approximately 3.1 Å, a smaller distance than the interlayer separation in graphite (3.45 Å); while electrons remain largely localized to individual C_{60} molecules, finite intermolecular overlap will exist.

6.7.2 Superconducting Fullerides

Solid C_{60} becomes superconducting when it is intercalated with three alkali metal atoms: K_3C_{60} becomes superconducting at temperatures below 18 K and Rb_3C_{60} at temperatures below 28 K. The alkali metal atoms occupy interstitial sites in both cases, and it is accepted that they act principally as electron donors into the mostly rigid LUMO of C_{60} (Cohen, 1993). Photoemission experiments clearly show the appearance of a new band near the Fermi level upon doping, with signal intensity proportional to doping level (Takahashi, 1993). Available experimental evidence and ab initio modelling both indicate that superconduc-

---

[6]The large mass of C_{60}, 712 atomic mass units, will promote low-energy acoustic vibrations in the solid.

tive pairing is mediated by vibrational modes on the C_{60} molecule, rather than by intermolecular modes or modes involving the alkali metal atoms (Schlüter et al., 1993). Two A_g modes and the eight Jahn-Teller-active H_g modes of C_{60} possess the correct symmetry to mediate electron-phonon coupling within a BCS pairing model, which is the accepted framework for superconductivity in the fullerides. Choices of the frequency range important to the coupling have varied among researchers,[7] but the experimental evidence supports the view that the H_g modes are directly involved. This is backed by isotope shift and Raman scattering data, a detailed review of which can be found in Schlüter et al. (1992b). Jishi and Dresselhaus (1992) have emphasized modes in the lower range of H_g frequencies, a frequency range that promotes an accurate theoretical fit to the temperature-dependent resistivity data (Cohen 1993). This range of frequencies is strongly Jahn-Teller coupled to electrons in the T_{1u} state of C_{60}, as calculations by de Coulon et al. (1992) have demonstrated. In support of this view are the calculations by Negri et al. (1988), which find that the 272 cm^{-1} H_g mode is the mode most actively coupled to a T_{1u} electron. This mode represents a spheroidal distortion of the C_{60} molecule, a distortion consistent with that predicted for the $T \otimes h$ ground state in Chapter 3.

6.7.3 $p^3 \otimes h$

The putative charge state of C_{60} in the fullerides is -3, and the unperturbed system is thus the $p^3 \otimes h$ case discussed in Section 3.5.2. Jahn-Teller coupling is limited to the low spin ^2D and ^2P states, and a set of pseudo-rotational eigenstates arise on the lowest APES (see Equations 3.69 and 3.70). In the absence of significant symmetry-lowering distortions, the molecule will pseudo-rotate through a series of distortions described by (3.70). In the fullerides, superconducting correlations are supposed to occur through coupling between the conduction electrons induced by intercalation and Jahn-Teller-type intramolecular modes. Correlation lengths vary from near 50 Å for Rb_3C_{60} to about 100 Å in K_3C_{60}, distances that range across 5 to 10 C_{60} molecules, and the pairing state is a BCS singlet. A real space picture of this situation is of limited help, but we can see the local stimulus: a low-spin electronic state (with character near that of a free-spin electron) interacts with the local "lattice" (the C_{60} molecule) by way of a local vibration. The $p^n \otimes h$ ground state is thus a "dressed state" for one of the electron spins involved in pairing. If we accept that two spin-$\frac{1}{2}$ states are involved in the BCS spin-singlet state, then it would appear that the Jahn-Teller ^2D state or ^2P state must be involved (rather than the ^4S state) if Jahn-Teller coupling is important. This means that the ^4S cannot be the local

---

[7]The mean vibrational frequency chosen by Jishi and Dresselhaus (1992) is about 348 cm^{-1}; Schlüter et al. (1992) choose frequencies in the range near 718 cm^{-1} and emphasize a broad range of H_g modes; Varma et al. (1991) suggest that high-frequency H_g modes are important and use an average phonon frequency of about 1450 cm^{-1}.

ground state on C_{60}. One result of this requirement can be seen by referring to Figures 3.8 and 3.9: any term splitting energy must be small in comparison with the Jahn-Teller energy. This, in fact, is also the expectation based on the simulations of Fagerström and Stafström (1993), who find that configuration interaction is a relatively small correction, both for C_{60}^{3-} and $X^{3+}C_{60}^{3-}$ complexes. Their calculations are consistent with the $p^3 \otimes h$ distortions described by (3.70). The calculations of Fagerström and Stafström also support the view that the minimum APES in $p^3 \otimes h$ will be only slightly warped by the presence of the alkali-earth ions, preserving the dynamic nature of the distortions. Several researchers have studied $p^3 \otimes h$ with regard to the suitability of C_{60}^{3+} as a negative-U center (used in the sense of the on-site correlation energy of a Hubbard Hamiltonian). Auerbach, Manini, and Tosatti (1994) and Manini, Tosatti, and Auerbach (1994) have discussed the general problem of vibronic coupling in C_{60}^{n-} at some length. O'Brien (1996) has also considered the effects of eight h_g modes, of configuration interaction, and of APES warping for Jahn-Teller couplings in these complexes.

Appendixes

A

Adiabatic Approximation

The adiabatic approximation to the molecular Schrödinger equation has several levels of complexity but is, in essence, an approximate method of solution that separates the electronic and nuclear motions. Historically, the physical view underlying this separation has been based on the difference between electronic and nuclear masses: the electrons, relatively light compared to the nuclei, move on a far faster time scale than do the nuclei, and so far as the nuclei are concerned, the electrons behave as a gas. It is the energy of this electron gas which then provides the potential energy for the nuclear motion (Fischer, 1989). Some, such as Essén (1977), have argued that a better theoretical justification for electronic-nuclear separability arises from the ability to partition the molecular potential energy into internal and relative parts. Whichever view one chooses, the experimental stimulus for making such a separation was recognized early in the twentieth century when it was discovered that molecular energy levels can be organized in hierarchies: molecular energy levels normally are such that they can be identified as belonging to rotational sequences of levels, each of which belongs to a given term in a vibrational sequence, with each of the vibrational sequences, in turn, associated with an electronic energy. These observations show that most molecules can be described as having a rotating, vibrating, semi-rigid system of nuclei (Essén, 1977; Kronig, 1930). The standard, and in many ways clearest, procedure for what has come to be known as the adiabatic approximation is due to Born and Huang (1954). They start with the time-independent Schrödinger equation,

$$H(r, Q) \, \Psi_n(r, Q) = \varepsilon_n \, \Psi_n(r, Q), \tag{A.1}$$

where $\Psi_n(r, Q)$ is the vibronic molecular wave function of energy ε_n (indexed by n), and r and Q represent sets of electronic and nuclear coordinates, respectively. The molecular Hamiltonian, $H(r, Q)$, is given by

$$H(r, Q) = T(r) + T(Q) + U(r, Q), \tag{A.2}$$

where the electronic and nuclear kinetic energies, $T(r)$ and $T(Q)$, are given by the usual expressions,

$$T(r) = -\frac{\hbar^2}{2m} \sum_i \nabla^2_{r_i}, \text{ etc.,} \tag{A.3}$$

and $U(r, Q)$ includes nuclear and electronic potential energies and electron-nuclei interactions. It is the last of these terms, the *vibronic* interactions, which concerns us. These vibronic terms make $U(r, Q)$ an operator in both the electronic and vibrational spaces. In the treatment of Born and Huang (1954), the molecular wavefunctions are expanded in a complete set of wave functions $u_{\Gamma_i}(r, Q)$,

$$\Psi_n(r, Q) = \sum_{\Gamma_i} \psi_{n, \Gamma_i}(Q) u_{\Gamma_i}(r, Q), \tag{A.4}$$

where Γ_i indexes the set of wavefunctions. The $u_{\Gamma_i}(r, Q)$ are selected by virtue of being eigenfunctions of the electronic Schrödinger equation,

$$[T(r) + U(r, Q)] u_{\Gamma_i}(r, Q) = E_\Gamma(Q) u_{\Gamma_i}(r, Q), \tag{A.5}$$

in which Q plays the part of a parameter: the electrons are assumed to be sensitive to different nuclear positions but relatively insensitive to the nuclear momenta in $T(Q)$. Even though exact solutions to (A.5) might not be possible in practice, the hierarchy of energies mentioned earlier opens the way for an approximation. It begins with the observation that the electronic energies for a fixed nuclear configuration (Q_o), given by the eigenvalues of $T(r) + U(r, Q_o)$, will typically be much larger than the energy corrections introduced by the remaining terms in (A.5). We can treat these other terms as a perturbation, $\{\ \}$, to the dominant electronic terms [ ], thus separating the Hamiltonian:

$$[T(r) + U(r, Q_o)] + \{U(Q) + M(r, Q)\}. \tag{A.6}$$

Here the perturbing terms in $U(r, Q)$ have been divided between a nuclear potential energy term, $U(Q)$, and a vibronic interaction, $M(r, Q)$. We should also note that we now are using the coordinates $\{Q\}$ to mean something more specific than we have up to this point: If Q_o represents the high-symmetry configuration of the molecule, then Q will represent a set of molecular normal modes that distort the molecule. The eigenvectors of $T(r) + U(r, Q_o)$ provide an electronic basis for a matrix representation of (A.5). We will denote the basis set by $\{u_{\Gamma_i}(r, Q_o)\}$, with these states eigenvectors of the equation

$$[T(r) + U(r, Q_o)] u_{\Gamma_i}(r, Q_o) = E_{o\Gamma} u_{\Gamma_i}(r, Q_o). \tag{A.7}$$

Γ will typically identify an electronic irreducible representation (irrep), and i will index the components of this irrep. The vibronic interaction matrices listed in Appendix E are represented in terms of the $\{u_{\Gamma_i}(r, Q_o)\}$ basis vectors, and each chapter's analysis starts from these states: for example, in Chapter 3 they are given a practical representation as $|\xi\rangle$, $|\eta\rangle$, and $|\zeta\rangle$, the three components of the electronic T_1 state. In this case, Γ is identified as T_1, one of the three-dimensional irreps of the icosahedral group, and i ranges over the labels $\{\xi, \eta, \zeta\}$. In our calculations, we will typically define energies relative to the $E_{o\Gamma}$. The Q_o configuration will, of course, be the icosahedral symmetry configuration. Corrections to the electronic energies ($E_{o\Gamma}$) arise from nuclear distortions to lower symmetry configurations ($Q \neq Q_o$) and from coupling between the normal modes (Q) and the electronic states ($u_{\Gamma_i}(r, Q_o)$). Solving the energy perturbation equation

$$[U(Q) + M(r, Q)] u_{\Gamma_i}(r, Q) = E_{\Gamma_i}(Q) u_{\Gamma_i}(r, Q) \tag{A.8}$$

will allow us to determine the energy corrections due to nuclear distortions and to vibronic coupling. (Notice that we have now included i as a subscript on the energy eigenvalue: the vibronic interaction will often lift the electronic degeneracy, and we must now include i as a label.) In Equation (A.8), $U(Q)$ typically will be the nuclear potential energy $\frac{1}{2}Q^2$ and $M(r, Q)$ will be the Jahn-Teller interaction. The solutions to (A.8), $u_{\Gamma_i}(r, Q)$, will be linear combinations of the $\{u_{\Gamma_i}(r, Q_o)\}$ basis vectors weighted by Q-dependent coefficients,

$$u_{\Gamma_j}(r, Q) = \sum_i c_i(Q) u_{\Gamma_i}(r, Q_o). \tag{A.9}$$

The states $\{u_{\Gamma_i}(r, Q_o)\}$, eigenvectors of the Jahn-Teller interaction $M(r, Q)$, are sometimes called static Jahn-Teller eigenvectors or adiabatic eigenvectors. The energy eigenvalues of (A.8), $\{E_{\Gamma_i}(Q)\}$, are the adiabatic potential energy surfaces (APES) mentioned in the chapters. Born and Huang (1954) assume that the APES are well separated so that coupling between them can be ignored. In this case, the vibronic wave function takes the approximate form of a so-called Born-Oppenheimer product of states,

$$\Psi_{\Gamma_i}(r, Q) = \psi_{\Gamma_i}(Q) u_{\Gamma_i}(r, Q). \tag{A.10}$$

$\psi_{\Gamma_i}(Q)$ describes the nuclear motion of the molecule when the electronic state of the molecule is associated with the APES defined by $E_{\Gamma_i}(Q)$. Wave functions such as (A.10) will usually be sufficient to describe the low-energy states of the systems we deal with, and we have used Born-Oppenheimer products for the various ground states discussed from Chapter 3 onwards. In cases where the $\{u_{\Gamma_i}(r, Q_o)\}$ are degenerate in energy, linear combinations of Born-Oppenheimer products will be necessary. The approximation procedure which has led to (A.10) is called the adiabatic approximation or the adiabatic Born-Oppenheimer approximation.

A.1 CORRECTIONS TO THE ADIABATIC APPROXIMATION

When (A.10) is substituted into the Schrödinger equation, the following equation results after applying closure:

$$\left[\left(E_\Gamma(Q) - \tfrac{1}{2} \nabla_Q^2 \right) \delta_{u',u} - \left(\tfrac{1}{2} \langle u' | \nabla_Q^2 | u \rangle + \langle u' | \nabla_Q | u \rangle \cdot \nabla_Q \right) \right] \psi_u(Q)$$

$$= \delta_{u',u} \, \mathcal{E}(Q) \, \psi_u(Q), \tag{A.11}$$

using the notational abbreviation $|u\rangle \equiv u_{\Gamma_i}(r, Q)$ for simplicity. The Kronecker delta arises due to the assummed orthonormality of the $\{|u\rangle\}$. As is evident, Equation (A.11) is a matrix equation with off-diagonal elements. The matrix operators involving ∇ and ∇^2 can have terms which couple the motions on the various APESs. Unitary operator techniques exist for including these in cases when APESs possess continuous symmetries (see the appendix on unitary transformations). Among the terms in (A.11), $\langle u |\nabla| u \rangle$ is of particular interest, since it can give rise to what has become known as the Berry phase (discussed in Chapter 1) when $|u\rangle$ is complex. A brief but useful review of different approaches to the adiabatic approximation is contained in Fischer (1989).

B

Quantum Tunneling Energies

B.1 INTRODUCTION

In nearly all the Jahn-Teller systems considered here, the symmetry types and ordering of the low-lying energy levels depend on solving the Schrödinger equation in a potential energy containing a series of wells in a multidimensional space. In this section we shall discuss the method used, and show to what extent we can rely on it.

B.2 ONE-DIMENSIONAL POTENTIALS

We start by pointing out that in a one-dimensional potential well, the eigenvalues of the Schrödinger equation are always ordered by the number of nodes, or zeros in the wave function. The lowest energy has no nodes, apart from those at $\pm\infty$, the second energy level has one node, and so on. This can be seen by putting the equation for the wave function ψ in a potential $V(x)$ in the form

$$\frac{\frac{d^2\psi}{dx^2}}{\psi} = -2(E - V(x)). \tag{B.1}$$

The left-hand side is the curvature of the wave function at a point, normalised by its magnitude, and at a given point x this increases with the energy of the state, E. Consequently if two solutions corresponding to different energies are compared in the region $E > V(x)$, the one with the larger energy will be impelled towards $\psi = 0$ more vigorously as the value of x changes, and will change sign more frequently. In regions where $E < V(x)$, the inequality still works the same way: the lower energy state is repelled from the axis more effectively. Clearly the state of higher energy will end up with more nodes. This is illustrated in Figure B.1, which shows a double well potential with the two lowest wave functions sketched in the case where the barrier is high compared with the energies of the states. These wave functions might be regarded as bonding and anti-bonding orbitals, and the energy of the antibonding orbital has to be higher to produce the increased value of the relative curvature of the wave function. Another familiar example of this principle is that of free

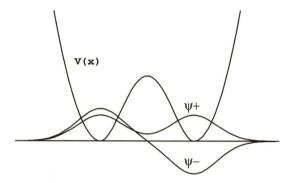

Figure B.1. The wave functions of the two states of lowest energy in a double well. Shows the bonding and antibonding type orbitals, ψ_+ and ψ_-.

waves, with $V(x) = 0$. Here a wave function such as $\sin kx$ has $E \propto k^2$ and a wavelength, $\lambda \propto 1/k$. The higher the energy, the more nodes per unit length. The example in Figure B.1 can also illustrate the way wave functions are built up out of components in the different wells when the barriers are high. In this approximation we take ψ_1 and ψ_2 to be normalised wave functions localised in the two wells, but each extending somewhat into the space between, and we take

$$\psi_+ = \frac{1}{\sqrt{2(1+S)}}(\psi_1 + \psi_2) \tag{B.2}$$

$$\psi_- = \frac{1}{\sqrt{2(1-S)}}(\psi_1 - \psi_2),$$

where S is the overlap between the two localised wave functions, appearing in the equations to ensure normalization. If the expectation value of the Hamiltonian, $\mathcal{H}$, is taken in these two states, the two energies come out as

$$E_+ = (\mathcal{H}_{11} + \mathcal{H}_{12})/(1+S) \tag{B.3}$$

$$E_- = (\mathcal{H}_{11} - \mathcal{H}_{12})/(1-S),$$

where $\mathcal{H}_{11}$ is the expectation value within either of the well states, and $\mathcal{H}_{12}$ is the cross term. At this point either the energies must be calculated numerically, or an asymptotic form can be found for the limit in which the overlap goes to zero. In that case a further expansion in small quantities gives

$$E_\pm = \mathcal{H}_{11} \pm \Delta, \quad \text{where} \quad \Delta = \mathcal{H}_{12} - S\mathcal{H}_{11}. \tag{B.4}$$

This is where the quantum-mechanical tunneling comes in. A process of matching wave functions across the boundaries of the wells and using the WKB approximation through the barrier gives

$$\Delta \propto - \exp(-I_T), \tag{B.5}$$

where I_T is the tunneling integral,

$$I_T = \int \sqrt{2(V - E)}\, dx \tag{B.6}$$

taken through the barrier. The sign puts the state with higher overlap and fewer nodes at the lower energy. This calculation is closely related to the more familiar problem of how long it will take a particle to leak through a barrier, and it is because of that application that I_T is known as the tunneling integral. The present application is a purely time-independent one; it was introduced for Jahn-Teller problems by Polinger (1974), but the above treatment, which is given in more detail later, follows O'Brien (1989). Before plunging into more dimensions, let us consider one more one-dimensional problem. Suppose that the configuration space is two-dimensional, but the potential has a deep circular trough, and that there are barriers regularly spaced along the trough that are still high, but much lower than the barrier confining the wave function to the trough. This can be treated approximately as a one-dimensional potential with periodic boundary conditions. To make the picture as simple as possible, we assume that there are four wells and four barriers. In the strong coupling approximation, and using the same notation as before, the low eigenstates are

$$\psi_a = \frac{1}{\sqrt{4(1+2S)}}(\psi_1 + \psi_2 + \psi_3 + \psi_4), \tag{B.7}$$

$$\psi_b = \frac{1}{\sqrt{4}}(\psi_1 + \psi_2 - \psi_3 - \psi_4),$$

$$\psi_c = \frac{1}{\sqrt{4}}(\psi_1 - \psi_2 - \psi_3 + \psi_4),$$

$$\psi_d = \frac{1}{\sqrt{4(1-2S)}}(\psi_1 - \psi_2 + \psi_3 - \psi_4),$$

and they are shown in Figure B.2. The number of nodes increases from ψ_a to ψ_d, as is reflected in the normalisation, and the asymptotic forms for the energies are consequently

$$E_a = \mathcal{H}_{11} + 2\Delta, \tag{B.8}$$

$$E_b = \mathcal{H}_{11},$$

$$E_c = \mathcal{H}_{11},$$

$$E_d = \mathcal{H}_{11} - 2\Delta.$$

Any further limitation, such as forbidding solutions of even-inversion symmetry would only cut out some of the states without altering the energies of the

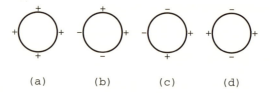

Figure B.2. The four wave functions for four wells on a circular trough.

others. When looking at more complicated systems of wells, it is convenient to introduce the overlap matrix. Here it would be

$$
\begin{bmatrix}
0 & S & 0 & S \\
S & 0 & S & 0 \\
0 & S & 0 & S \\
S & 0 & S & 0
\end{bmatrix}
\tag{B.9}
$$

where an S is entered for each pair of neighbouring wells. The eigenvalues of this matrix are $2S$, 0, 0, and $-2S$ and the eigenvectors are $(1, 1, 1, 1)$, $(1, 1, -1, -1)$, $(1, -1, -1, 1)$, and $(1, -1, 1, -1)$ respectively. Thus the eigenvalues and eigenvectors of the overlap matrix give the actual energies and wave functions directly, according to the rule that maximum overlap with fewest nodes corresponds to minimum energy. This treatment is clearly very similar to the tight binding approximation for electronic states in crystal.

B.3 HIGHER DIMENSIONALITY

With added dimensions everything becomes increasingly complicated. In 1937, Kapur and Peierls showed that the path to choose for the tunneling integral was one of minimum action, the same as a classical trajectory. Specifically, in terms of the WKB (Wentzel-Kramers-Brillouin) approximation,

$$
\psi(\mathbf{r}) = \exp(-S(\mathbf{r}),
\tag{B.10}
$$

if $S(\mathbf{r})$ is real, it satisfies the relation

$$
|S(\mathbf{r}_1) - S(\mathbf{r}_2)| = \min \int_{\mathbf{r}_1}^{\mathbf{r}_2} \sqrt{2(V - E)}\, ds,
\tag{B.11}
$$

where the path between $\mathbf{r}_1$ and $\mathbf{r}_2$ is chosen to minimize the integral. In the Jahn-Teller systems discussed here, such a path between minima will exist. It will be a compromise between the shortest distance and the path through a saddle point that minimises the peak value of $\sqrt{2(V - E)}$. The discussion of the ground state of these systems is based on the assumption that the effect on the energy of the overlap of functions in different wells is similar to that of a one-dimensional system, with the Kapur-Peierls integral playing the part of the tunneling integral, I_T. Symmetry always ensures that all the tunneling paths are equivalent. This assumption is what leads to the use of the overlap matrix to sort out the symmetry types and order of the ground states.

B.4 THE WKB APPROXIMATION AND ITS APPLICATION

In this section we discuss the theory underlying the use of the tunneling integral and the justification for its use. The WKB (Wentzel-Kramers-Brillouin) approximation is designed to be used in circumstances where the kinetic energy term in the Schrödinger equation is small compared with $|E - V|$. Here we only use it in the classically forbidden region where $V >> E$. The Schrödinger equation in the classically forbidden region is converted to the WKB form by making the substitution

$$\psi(\mathbf{r}) = \exp(-S(\mathbf{r})), \tag{B.12}$$

so that

$$\nabla^2 \psi = (|\nabla S|^2 - \nabla^2 S)\psi. \tag{B.13}$$

The Schrödinger equation

$$-\frac{\hbar^2}{2m}\nabla^2 \psi + V(\mathbf{r})\psi = E\psi \tag{B.14}$$

is thus replaced by

$$-\frac{\hbar^2}{2m}(|\nabla S|^2 - \nabla^2 S) + V(\mathbf{r}) = E, \tag{B.15}$$

where the factor $\hbar^2/m$ is retained temporarily to serve as an expansion parameter. This form is then used in circumstances when it can reasonably be assumed that S is large, so that

$$|\nabla S|^2 \gg \nabla^2 S. \tag{B.16}$$

S is then expanded in powers of $\hbar$, and the largest term is given by

$$|\nabla S_0|^2 = \frac{2m}{\hbar^2}(V - E) \tag{B.17}$$

and to the next order S_1 must satisfy

$$2\nabla S_1.\nabla S_0 - \nabla^2 S_0 = 0. \tag{B.18}$$

In the one-dimensional case it is possible to integrate these equations to get a solution in the form

$$
\begin{aligned}
S_0(x) &= \pm \int^x \sqrt{\tfrac{2m}{\hbar^2}(V(x) - E)}\, dx, \\
\psi(x) &\propto (V(x) - E)^{-\frac{1}{4}} \exp(-S_0(x)).
\end{aligned} \tag{B.19}
$$

This is a good approximation to the true solution when $(V(x) - E)$ is large. The discussion of the three-dimensional case by Kapur and Peierls gives the expression for $S_0(\mathbf{r})$. They do not go on to the next approximation, which gives the prefactor, but this can be found in some work reported by Ranfagni (1977), giving

$$
\begin{aligned}
\psi(\mathbf{s}) &\propto (\kappa \Sigma)^{-\frac{1}{2}} \exp(-\int_{\mathbf{s}} \kappa(s)\, ds), \\
\kappa &= \sqrt{\tfrac{2m}{\hbar^2}(V - E)}
\end{aligned} \tag{B.20}
$$

where $\mathbf{s}$ is a classical trajectory at each point, and Σ is the normal surface area of a flux tube at each point.

B.4.1 One-Dimensional Application

The problem to be solved is shown in Figure B.3, where a double well potential is shown with harmonic oscillator potentials fitted to the two wells. The wave function ψ_1 centered on the first well consists mostly of the lowest harmonic oscillator wave function in that well. This wave function is a solution of the Schrödinger equation wherever $V(x)$ coincides with the harmonic oscillator potential, but under the barrier $V(x)$ is lower, and in this region a WKB solution is used. The two solutions are matched in slope and amplitude at a point that is far enough under the barrier for the WKB solution to hold good, but before the harmonic oscillator potential has parted company from $V(x)$. The coordinate of the matching point does not appear in the final answer; it is simply a condition of validity that such a point exists. The process of matching chooses the WKB wave function that decreases away from the well, as is shown in Figure B.3. This wave function cannot be matched to a wave function in the second well; instead it can be assumed to be terminated smoothly by a matched function that falls off exponentially with increasing x. An exactly similar process gives ψ_2, the wave function centered in the second well. The two non-orthogonal wave functions are then used as the bases to find explicit expressions for the quantities $\mathcal{H}_{11}$, $\mathcal{H}_{12}$, and S that enter into the expressions for the energies (B.3). The final result of all the algebra and asymptotic approximations gives

$$
\Delta = \mathcal{H}_{12} - S\mathcal{H}_{11} = -\left(\frac{1}{4\pi}\right)^{1/2} \exp\left(-I_T - \frac{1}{2}\right), \tag{B.21}
$$

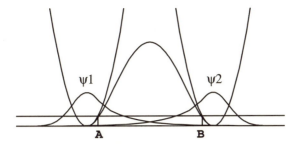

Figure B.3. Sketch of this one-dimensional tunneling problem, showing harmonic oscillator potentials fitted to the potential in each well and the wave functions extended into the barrier region by the WKB approximation. A and B mark the intersection of the ground harmonic oscillator energy level with the potential. The range of the tunneling integral is from A to B.

where the tunneling integral I_T is taken right through the classically inaccessible region between A, and B on Figure B.3, that is,

$$I_T = \int_A^B \sqrt{2(V - E)}\, dx. \qquad (B.22)$$

B.4.2 WKB in More Dimensions

A similar construction for wave functions centered on many dimensional wells can be set up, each wave function consisting of a ground state harmonic oscillator wave function in the well, a matched decaying WKB wave function under the barrier, and a matched decaying part near the next well. A result analogous to (B.21) in three dimensions is

$$\Delta_3 = \left(\frac{3}{4\pi e}\right)^{3/2} \int \exp(-I_T)\, d\Omega, \qquad (B.23)$$

where $d\Omega$ is an element of solid angle subtended at the centre of a well. The surface integral appears because the matching must take place over a surface, and the tunneling integral is inside the surface integral because its value depends on the particular choice of ends. In fact the only significant contribution will come from a pencil of trajectories surrounding the best path, but it is difficult to see how to choose the area of the pencil. However unsatisfactory, this formal result does at least support the use of the overlap matrix to give the relative energies of the ground states. Other approaches to multidimensional tunneling are developed by Huang et al. (1990) and by Auerbach and Kivelson (1985), and more work on the Jahn-Teller problem with these approaches in view would be desirable.

C

E ⊗ ε

The Hamiltonian of an electronic doublet linearly coupled to two vibrational modes in cubic symmetry is only slightly more complicated than the $E \otimes \beta_{1g}$ example outlined in Chapter 1. However, the $E \otimes \epsilon$ system is much richer in the phenomena it gives rise to, since it possesses a minimum APES that is a continuum in phase space with SO(2) symmetry. The interwoven simplicity and complexity of this system has made it a useful reference case when discussing various icosahedral systems, and we have referred to it in our earlier discussions of $T \otimes h$ and $H \otimes (g \oplus h)$. Englman (1972) and Bersuker and Polinger (1989) are standard references for the $E \otimes \epsilon$ system.

The matrix Hamiltonian for the linear interaction in the space of the electronic doublet $\{| \theta \rangle, | \epsilon \rangle \}$ takes the form

$$\mathcal{H} = -\frac{1}{2} \left(\frac{\partial^2}{\partial Q_\theta^2} + \frac{\partial^2}{\partial Q_\epsilon^2} + Q_\theta^2 + Q_\epsilon^2 \right) + k_\epsilon^E \begin{bmatrix} -Q_\theta & Q_\epsilon \\ Q_\epsilon & Q_\theta \end{bmatrix}. \tag{C.1}$$

The SO(2) symmetry of the energy minimum on the lower of the two APESs is best expressed through a change to variables $\{Q, \phi\}$: $Q_\theta = Q \cos \phi$ and $Q_\epsilon = Q \sin \phi$. Within this new parametrization, the interaction matrix can be diagonalized by a transformation that is simpler than but analogous to those used in $T \otimes h$ and $H \otimes h_2$. The transformation

$$\mathcal{H}' = T^{-1} \mathcal{H} T$$

$$= -\frac{1}{2} \left(\frac{\partial^2}{\partial Q^2} + \frac{1}{Q} \frac{\partial}{\partial Q} \right) - \frac{1}{2Q^2} \begin{bmatrix} -\frac{1}{4} + \frac{\partial^2}{\partial \phi^2} & \frac{\partial}{\partial \phi} \\ -\frac{\partial}{\partial \phi} & -\frac{1}{4} - \frac{\partial^2}{\partial \phi^2} \end{bmatrix} + \frac{1}{2} Q^2$$

$$+ k_\epsilon^E Q \begin{bmatrix} -1 & 0 \\ 0 & 1 \end{bmatrix}, \tag{C.2}$$

where

$$T = \begin{bmatrix} \cos(\phi/2) & \sin(\phi/2) \\ -\sin(\phi/2) & \cos(\phi/2) \end{bmatrix}, \tag{C.3}$$

represents a rotation in the $\{Q_\theta, Q_\epsilon\}$ phase space through an angle $\phi/2$. The new basis vectors for the diagonalized interaction are

$$| u_1 \rangle = \cos(\phi/2) | \theta \rangle - \sin(\phi/2) | \epsilon \rangle \tag{C.4}$$

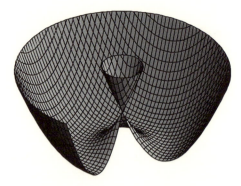

Figure C.1. The lower and upper APESs for the linear E⊗ε system. The $\{Q_\theta, Q_\epsilon\}$ plane is parallel to the circular minimum trough and has its origin at the conical intersection.

and

$$| u_2 \rangle = \sin(\phi/2) | \theta \rangle + \cos(\phi/2) | \epsilon \rangle. \tag{C.5}$$

The two APESs are shown in Figure C.1. The eigenvector $| u_1 \rangle$ is associated with the lower APES, $U_1 = \frac{1}{2}Q^2 - k_\epsilon^E Q$, which reaches a minimum of $-\frac{1}{2}(k_\epsilon^E)^2$ for $Q = k$. The off-diagonal terms in (C.2), which arise from the kinetic matrix elements $\langle u | \nabla^2 | u \rangle$, will connect motions on the two APESs, but we can neglect their effect in determining the strong coupling ($k_\epsilon^E \gg 1$) ground state. Within the Born-Oppenheimer approximation, this state takes the form,

$$\Psi = e^{(im\phi)} f(Q) | u_1 \rangle. \tag{C.6}$$

While the normal mode coordinates are invariant under the mapping $\phi \rightarrow \phi + 2\pi$, the electronic eigenvector $| u_1 \rangle$ transforms to the negative of itself, $| u_1 \rangle \rightarrow - | u_1 \rangle$. For Ψ to be single-valued, m must be half-odd integral: $m = \pm\frac{1}{2}, \pm\frac{3}{2}, \dots$. In the language of the Berry phase, the half-integral requirement is equivalent to the appearance of a π phase change in $| u_1 \rangle$ as it is adiabatically transported along the minimum energy trough of U_1, a path that encircles a degeneracy.

Placing (C.6) in (C.2) and performing a separation of variables leads to low-energy states of the form

$$\Psi = Q^{-\frac{1}{2}} \chi_n(Q - k) e^{(im\phi)} | u_1 \rangle, \tag{C.7}$$

where the $\chi_n(Q - k)$ is the nth oscillator wave function for oscillations against the sides of the trough. Pseudo-rotational motions along the potential energy trough have energies that go as

$$E \approx \frac{1}{2} - \frac{1}{2}(k_\epsilon^E)^2 + m^2/2(k_\epsilon^E)^2, \tag{C.8}$$

where $\frac{1}{2}$ is the zero point oscillation energy. The appearance of m^2 shows that the E $\otimes$ ε ground state is a doublet labelled by $m = \pm\frac{1}{2}$.

The artificially high SO(2) symmetry of the ground state can be broken by including a warping term in the Hamiltonian, such as $K Q^3 \cos(3\phi)$, the A$_1$ cubic invariant of the group O$_h$. Including this term produces three wells along the trough where the wave function can be localized. For large warpings, the system will be localized in one of the wells, producing a static distortion, while for smaller warpings, the wave function will tunnel between wells.

The effects of tunneling on the wave function can be addressed by the methods of Appendix B, and the tunneling matrix can be set up in the basis of states provided by the localized wave functions corresponding to each well,

$$\begin{bmatrix} 0 & -S & -S \\ -S & 0 & -S \\ -S & -S & 0 \end{bmatrix}. \tag{C.9}$$

Here the minus signs occur due to the requirement that

$$\Psi = \psi(Q, \phi) \mid u_1\rangle$$

be single-valued, and this means that ψ must change sign between every pair of minima. The matrix has a single root at $-3S$ and a twofold degenerate root at S, which implies a doubly-degenerate ground state, since energy decreases with increasing wave function overlap.

D

The Group I

TABLE D.1
Character Table for I_h
In the table $r = (1 + \sqrt{5})/2$; r is the golden ratio, satisfying the equation $r^2 = r + 1$.

I_h	E	$12C_5$	$12C_5^2$	$20C_3$	$15C_2$	P	$12S_{10}^3$	$12S_{10}$	$20S_3$	$15\sigma_v$
A_g	+1	+1	+1	+1	+1	+1	+1	+1	+1	+1
T_{1g}	+3	r	$1-r$	0	−1	+3	r	$1-r$	0	−1
T_{2g}	+3	$1-r$	r	0	−1	+3	$1-r$	r	0	−1
G_g	+4	−1	−1	+1	0	+4	−1	−1	+1	0
H_g	+5	0	0	−1	+1	+5	0	0	−1	+1
A_u	+1	+1	+1	+1	+1	−1	−1	−1	−1	−1
T_{1u}	+3	r	$1-r$	0	−1	−3	$-r$	$r-1$	0	+1
T_{2u}	+3	$1-r$	r	0	−1	−3	$r-1$	$-r$	0	+1
G_u	+4	−1	−1	+1	0	−4	+1	+1	−1	0
H_u	+5	0	0	−1	+1	−5	0	0	+1	−1

Source: Dresselhaus, Dresselhaus, and Eklund (1992).

TABLE D.2

Character Table for the I Double Group

In the table $r = (1 + \sqrt{5})/2$; r is the golden ratio, satisfying the equation $r^2 = r + 1$.

I	E	$12C_5$	$12C_5^2$	$20C_3$	$15C_2$	R	$12C_5^4$	$12C_{5^3}$	$20C_3^2$
A	+1	+1	+1	+1	+1	+1	+1	+1	+1
T_1	+3	r	$1-r$	0	-1	+3	r	$1-r$	0
T_2	+3	$1-r$	r	0	-1	+3	$1-r$	r	0
G	+4	-1	-1	+1	0	+4	-1	-1	+1
H	+5	0	0	-1	+1	+5	0	0	-1
Γ_6	+2	r	$r-1$	+1	0	-2	$-r$	$1-r$	-1
Γ_7	+2	$1-r$	$-r$	1	0	-2	$r-1$	r	-1
Γ_8	+4	1	-1	-1	0	-4	-1	1	1
Γ_9	+6	-1	1	0	0	-6	+1	-1	0

Source: Herzberg (1966).

TABLE D.3

Direct Product Representations for the Double Group I

	A	T_1	T_2	G	H	Γ_6	Γ_7	Γ_8	Γ_9
A	A	T_1	T_2	G	H	Γ_6	Γ_7	Γ_8	Γ_9
T_1	[A, H], {T_1}	G, H	T_2, G, H	T_1, T_2, G, H	Γ_6, Γ_8	Γ_9	Γ_6, Γ_8, Γ_9	Γ_7, Γ_8, $2\Gamma_9$
T_2		[A, H], {T_2}	T_1, G, H	T_1, T_2, G, H	Γ_9	Γ_7, Γ_8	Γ_7, Γ_8, Γ_9	Γ_6, Γ_8, $2\Gamma_9$
G			[A, G, H], {T_1, T_2}	T_1, T_2, G, 2H	Γ_7, Γ_9	Γ_6, Γ_9	Γ_8, $2\Gamma_9$	Γ_6, Γ_7, $2\Gamma_8$, $2\Gamma_9$
H				[A, G, 2H], {T_1, T_2, G}	Γ_8, Γ_9	Γ_8, Γ_9	Γ_6, Γ_7, Γ_8, $2\Gamma_9$	Γ_6, Γ_7, $2\Gamma_8$, $3\Gamma_9$
Γ_6						[T_1], {A}	G	T_1, H	T_2, G, H
Γ_7							[T_2], {A}	T_2, H	T_1, G, H
Γ_8								[T_1, T_2, G], {A, H}	T_1, T_2, 2G, 2H
Γ_9									[$2T_1$, $2T_2$, G, H], {A, G, 2H}

Notes: Symmetric squares are in [ ], antisymmetric squares in {}.

E

Jahn-Teller Interaction Matrices and Their Bases

E.1 BASIS STATES

In Chapter 2, Table 2.2 lists the classification of spherical harmonics in terms of the irreps of I_h, and we shall use that list to provide explicit bases for each irrep. The definition and numbering of the bases will define the choice of bases and the order in which they appear in interaction matrices throughout this book. The basis functions will be listed without overall normalization, but consistently normalized within each set.

E.1.1 $L = 1$ and $L = 2$ Bases

Both these angular momentum states go over into complete irreps, so all that is required is to define the numbering. The T_{1u} or p bases are

$$
\begin{aligned}
T_{1u}(L = 1): & \\
p_1 &= x \\
p_2 &= y \\
p_3 &= z.
\end{aligned}
\tag{E.1}
$$

The H_g or d bases will be identified here by the extra index, 2, that indicates they are derived from $L = 2$. This is because it is also sometimes necessary to use a set of H bases derived from $L = 4$, and the index can be dropped where the distinction is not needed. The bases as we use them are

$$
\begin{aligned}
H_{2g}(L = 2): & & & \\
h_1 &= \tfrac{1}{2}(2z^2 - x^2 - y^2) &=& \tfrac{1}{2}(3\cos^2\theta - 1) \\
h_2 &= \sqrt{3}xz &=& \tfrac{\sqrt{3}}{2}\sin 2\theta \cos\phi \\
h_3 &= \sqrt{3}xy &=& \tfrac{\sqrt{3}}{2}\sin^2\theta \sin 2\phi \\
h_4 &= \tfrac{\sqrt{3}}{2}(x^2 - y^2) &=& \tfrac{\sqrt{3}}{2}\sin^2\theta \cos 2\phi \\
h_5 &= \sqrt{3}yz &=& \tfrac{\sqrt{3}}{2}\sin 2\theta \sin\phi.
\end{aligned}
\tag{E.2}
$$

E.1.2 Bases from $L = 3$ and Upwards

States from $L = 3$ and up split up in I_h, as listed in Table 2.4. One way of separating them out is to diagonalize a crystal field of icosahedral symmetry in the angular momentum states. The form of the field used here is

$$V_{icos} = 231z^6 - 315r^2z^4 + 105r^4z^2 - 5r^6 + 42z(x^5 - 10x^3y^2 + 5xy^4). \quad \text{(E.3)}$$

This field is written with the z-axis along a fivefold axis of the icosahedron, and the positive quadrant of the $y = 0$ plane containing another fivefold axis, so that will be the choice of orientation for all the basis states given here. A collection of basis states produced by this process is shown below; the normalization is arbitrary and only consistent within each set. Where the states are given in terms of the angles (θ, ϕ), these are spherical polar coordinates corresponding to $r = 1$.

$T_{2u}(L = 3)$:
$$
\begin{aligned}
t_1 &= 3x^2y - y^3 + 6xyz &&= \sin^3\theta \cos 3\phi + 3\cos\theta \sin^2\theta \cos 2\phi \\
t_2 &= 3xy^2 - x^3 + 3z(x^2 - y^2) &&= -\sin^3\theta \sin 3\phi + 3\cos\theta \sin^2\theta \sin 2\phi \\
t_3 &= 5z^3 - 3zr^2 &&= 5\cos^3\theta - 3\cos\theta
\end{aligned}
$$
$$\text{(E.4)}$$

$G_u(L = 3)$:
$$
\begin{aligned}
g_1 &= y^3 - 3x^2y - 4xyz &&= -\sin^3\theta \sin 3\phi - 2\cos\theta \sin^2\theta \sin 2\phi, \\
g_2 &= x^3 - 3xy^2 - 2z(x^2 - y^2) &&= \sin^3\theta \cos 3\phi - 2\cos\theta \sin^2\theta \cos 2\phi, \\
g_3 &= y(r^2 - 5z^2) &&= \sin\theta(1 - 5\cos^2\theta)\sin\phi, \\
g_4 &= x(5z^2 - r^2) &&= -\sin\theta(1 - 5\cos^2\theta)\cos\phi.
\end{aligned}
$$
$$\text{(E.5)}$$

$G_g(L = 4)$:
$$
\begin{aligned}
g_1 &= (x^2 - y^2)(x^2 + y^2 - 6z^2) - z(3y^2x - x^3), \\
g_2 &= -2xy(x^2 + y^2 - 6z^2) + z(3x^2y - y^3), \\
g_3 &= -xz(3x^2 + 3y^2 - 4z^2) + x^4 + y^4 - 6x^2y^2, \\
g_4 &= -yz(3x^2 + 3y^2 - 4z^2) - 4yx(x^2 - y^2).
\end{aligned}
$$
$$\text{(E.6)}$$

$H_{4g}(L = 4)$:
$$
\begin{aligned}
h_1 &= (3 - 30\cos^2\theta + 35\cos^4\theta)\sqrt{3}, \\
h_2 &= (12\sin 2\theta - 28\cos^2\theta \sin 2\theta)\cos\phi + 7\sin^4\theta \cos 4\phi, \\
h_3 &= (-2 + 16\cos^2\theta - 14\cos^4\theta)\sin 2\phi - 14\sin^2\theta \sin 2\theta \sin 3\phi, \\
h_4 &= (-2 + 16\cos^2\theta - 14\cos^4\theta)\cos 2\phi + 14\sin^2\theta \sin 2\theta \cos 3\phi, \\
h_5 &= (12\sin 2\theta - 28\cos^2\theta \sin 2\theta)\sin\phi - 7\sin^4\theta \sin 4\phi.
\end{aligned}
$$
$$\text{(E.7)}$$

$H_u(L = 5)$:
$$
\begin{aligned}
h_1 &= \sqrt{3}(1 - 2\cos^2\theta + \cos^4\theta)\sin\theta\sin 5\phi, \\
h_2 &= -(1 - 14\cos^2\theta + 21\cos^4\theta)\sin\theta\sin\phi \\
    &\quad +3(\cos\theta - 2\cos^3\theta + \cos^5\theta)\sin 4\phi, \\
h_3 &= 4(\cos\theta - 4\cos^3\theta + 3\cos^5\theta)\cos 2\phi \\
    &\quad +(1 - 10\cos^2\theta + 9\cos^4\theta)\sin\theta\cos 3\phi, \\
h_4 &= -4(\cos\theta - 4\cos^3\theta + 3\cos^5\theta)\sin 2\phi \\
    &\quad +(1 - 10\cos^2\theta + 9\cos^4\theta)\sin\theta\sin 3\phi, \\
h_5 &= (1 - 14\cos^2\theta + 21\cos^4\theta)\sin\theta\cos\phi \\
    &\quad +3(\cos\theta - 2\cos^3\theta + \cos^5\theta)\cos 4\phi.
\end{aligned}
\tag{E.8}
$$

$T_{1g}(L = 6)$:
$$
\begin{aligned}
t_1 &= (10\cos\theta - 60\cos^3\theta + 66\cos^5\theta)\sin\theta\sin\phi \\
    &\quad +(-1 + 13\cos^2\theta - 23\cos^4\theta + 11\cos^6\theta)\sin 4\phi \\
    &\quad +(1 - 3\cos^2\theta + 3\cos^4\theta - \cos^6\theta)\sin 6\phi, \\
t_2 &= -(10\cos\theta - 60\cos^3\theta + 66\cos^5\theta)\sin\theta\cos\phi \\
    &\quad +(-1 + 13\cos^2\theta - 23\cos^4\theta + 11\cos^6\theta)\cos 4\phi \\
    &\quad -(1 - 3\cos^2\theta + 3\cos^4\theta - \cos^6\theta)\cos 6\phi, \\
t_3 &= (-10\cos\theta + 20\cos^3\theta - 10\cos^5\theta)\sin\theta\sin 5\phi
\end{aligned}
\tag{E.9}
$$

E.1.3 Interaction Matrices

In terms of the basis states listed above, the interaction matrices are as listed below. As discussed in Chapter 2 (2.4), they have the general form

$$
\sum_{\Lambda,\lambda} Q^{\Lambda}_{\lambda}\, \mathbf{U}^{\Gamma}(\Lambda\lambda).
\tag{E.10}
$$

The normalization is arbitrary. The matrices are
for $T_1 \otimes h$:

$$
\frac{1}{2}
\begin{bmatrix}
Q_1 - \sqrt{3}Q_4 & -\sqrt{3}Q_3 & -\sqrt{3}Q_2 \\
-\sqrt{3}Q_3 & Q_1 + \sqrt{3}Q_4 & -\sqrt{3}Q_5 \\
-\sqrt{3}Q_2 & -\sqrt{3}Q_5 & -2Q_1
\end{bmatrix}
\tag{E.11}
$$

for $T_2 \otimes h$:

$$
\frac{1}{2}
\begin{bmatrix}
Q_1 - \sqrt{3}Q_2 & \sqrt{3}Q_5 & \sqrt{3}Q_4 \\
\sqrt{3}Q_5 & Q_1 + \sqrt{3}Q_2 & \sqrt{3}Q_3 \\
\sqrt{3}Q_4 & \sqrt{3}Q_3 & -2Q_1
\end{bmatrix}
\tag{E.12}
$$

for $G \otimes g$:

$$
\begin{bmatrix}
-g_3 & -g_4 & -g_1 + g_3 & -g_2 - g_4 \\
-g_4 & g_3 & g_2 - g_4 & -g_1 - g_3 \\
-g_1 + g_3 & g_2 - g_4 & g_1 & -g_2 \\
-g_2 - g_4 & -g_1 - g_3 & -g_2 & -g_1
\end{bmatrix}
\tag{E.13}
$$

for $G \otimes h$:

$$\begin{bmatrix} 2h_2 - \sqrt{3}h_1 & 2h_5 & h_2 - h_4 & h_3 - h_5 \\ 2h_5 & -\sqrt{3}h_1 - 2h_2 & -h_3 - h_5 & -h_2 - h_4 \\ h_2 - h_4 & -h_3 - h_5 & \sqrt{3}h_1 - 2h_4 & -2h_3 \\ h_3 - h_5 & -h_2 - h_4 & -2h_3 & \sqrt{3}h_1 + 2h_4 \end{bmatrix} \quad (E.14)$$

for $H \otimes g$:

$$\begin{bmatrix} 0 & 2\sqrt{3}g_3 & 2\sqrt{3}g_2 & -2\sqrt{3}g_1 & 2\sqrt{3}g_4 \\ 2\sqrt{3}g_3 & -4g_1 & g_2 - g_4 & g_1 - g_3 & 4g_2 \\ 2\sqrt{3}g_2 & g_2 - g_4 & -4g_3 & -4g_4 & -g_1 - g_3 \\ -2\sqrt{3}g_1 & g_1 - g_3 & -4g_4 & 4g_3 & g_2 + g_4 \\ 2\sqrt{3}g_4 & 4g_2 & -g_1 - g_3 & g_2 + g_4 & 4g_1 \end{bmatrix} \quad (E.15)$$

for $H \otimes h_2$:

$$\begin{bmatrix} 2h_1 & h_2 & -2h_3 & -2h_4 & h_5 \\ h_2 & h_1 + \sqrt{3}h_4 & \sqrt{3}h_5 & \sqrt{3}h_2 & \sqrt{3}h_3 \\ -2h_3 & \sqrt{3}h_5 & -2h_1 & 0 & \sqrt{3}h_2 \\ -2h_4 & \sqrt{3}h_2 & 0 & -2h_1 & -\sqrt{3}h_5 \\ h_5 & \sqrt{3}h_3 & \sqrt{3}h_2 & -\sqrt{3}h_5 & h_1 - \sqrt{3}h_4 \end{bmatrix} \quad (E.16)$$

for $H \otimes h_4$:

$$\begin{bmatrix} 6\sqrt{3}h_1 & -4\sqrt{3}h_2 & \sqrt{3}h_3 & \sqrt{3}h_4 & -4\sqrt{3}h_5 \\ -4\sqrt{3}h_2 & -4\sqrt{3}h_1 + 2h_4 & -7h_3 + 2h_5 & 2h_2 + 7h_4 & 2h_3 \\ \sqrt{3}h_3 & -7h_3 + 2h_5 & \sqrt{3}h_1 - 7h_2 & -7h_5 & 2h_2 - 7h_4 \\ \sqrt{3}h_4 & 2h_2 + 7h_4 & -7h_5 & \sqrt{3}h_1 + 7h_2 & -7h_3 - 2h_5 \\ -4\sqrt{3}h_5 & 2h_3 & 2h_2 - 7h_4 & -7h_3 - 2h_5 & -4\sqrt{3}h_1 - 2h_4 \end{bmatrix}$$
$$(E.17)$$

for $H \otimes h_a$:

$$\begin{bmatrix} 0 & \sqrt{3}h_2 & -\sqrt{3}h_3 & -\sqrt{3}h_4 & \sqrt{3}h_5 \\ \sqrt{3}h_2 & \sqrt{3}h_1 + h_4 & h_3 + h_5 & h_2 - h_4 & h_3 \\ -\sqrt{3}h_3 & h_3 + h_5 & -\sqrt{3}h_1 + h_2 & h_5 & h_2 + h_4 \\ -\sqrt{3}h_4 & h_2 - h_4 & h_5 & -\sqrt{3}h_1 - h_2 & h_3 - h_5 \\ \sqrt{3}h_5 & h_3 & h_2 + h_4 & h_3 - h_5 & \sqrt{3}h_1 - h_4 \end{bmatrix} \quad (E.18)$$

for $H \otimes h_b$:

$$\begin{bmatrix} \frac{4h_1}{\sqrt{3}} & -\frac{h_2}{\sqrt{3}} & -\frac{h_3}{\sqrt{3}} & -\frac{h_4}{\sqrt{3}} & -\frac{h_5}{\sqrt{3}} \\ -\frac{h_2}{\sqrt{3}} & -\frac{h_1}{\sqrt{3}} + h_4 & -h_3 + h_5 & h_2 + h_4 & h_3 \\ -\frac{h_3}{\sqrt{3}} & -h_3 + h_5 & -\frac{h_1}{\sqrt{3}} - h_2 & -h_5 & h_2 - h_4 \\ -\frac{h_4}{\sqrt{3}} & h_2 + h_4 & -h_5 & -\frac{h_1}{\sqrt{3}} + h_2 & -h_3 - h_5 \\ -\frac{h_5}{\sqrt{3}} & h_3 & h_2 - h_4 & -h_3 - h_5 & -\frac{h_1}{\sqrt{3}} - h_4 \end{bmatrix} \quad (E.19)$$

F

Transformations

F.1 PARAMETRIZATIONS OF THE h MODES

The choice of parameters for the h modes, (3.5), is based first on the observation that points on a three-dimensional sphere of radius Q (3.3) can be written in terms of the matrices

$$C_D(\theta) = \begin{bmatrix} \frac{1}{2}(3c^2 - 1) & -\sqrt{3}sc & 0 & \frac{\sqrt{3}}{2}s^2 & 0 \\ \sqrt{3}sc & (2c^2 - 1) & 0 & -sc & 0 \\ 0 & 0 & c & 0 & s \\ \frac{\sqrt{3}}{2}s^2 & sc & 0 & \frac{1}{2}(1+c^2) & 0 \\ 0 & 0 & -s & 0 & c \end{bmatrix}, \quad \text{(F.1)}$$

where $s = \sin\theta$ and $c = \cos\theta$, and

$$D_D(\phi) = \begin{bmatrix} 1 & 0 & 0 & 0 & 0 \\ 0 & \cos\phi & 0 & 0 & -\sin\phi \\ 0 & 0 & \cos 2\phi & \sin 2\phi & 0 \\ 0 & 0 & -\sin 2\phi & \cos 2\phi & 0 \\ 0 & \sin\phi & 0 & 0 & \cos\phi \end{bmatrix} \quad \text{(F.2)}$$

together with the vector

$$\mathbf{Q}_0 = \begin{bmatrix} Q \\ 0 \\ 0 \\ 0 \\ 0 \end{bmatrix} \quad \text{(F.3)}$$

in the form

$$\mathbf{Q}_{\text{sphere}} = \begin{bmatrix} Q_1 \\ Q_2 \\ Q_3 \\ Q_4 \\ Q_5 \end{bmatrix} = D_D \times C_D \times \mathbf{Q}_0. \quad \text{(F.4)}$$

This shows how a general point on the three-dimensional sphere can be generated by rotations from a special point. If we now take a general point P in the

$\{Q_i\}$ space and define

$$Q = \sqrt{Q_1^2 + Q_2^2 + Q_3^2 + Q_4^2 + Q_5^2},\qquad\text{(F.5)}$$

we can find a unique point on the three-dimensional sphere of radius Q that is nearer to P than any other. When that point is $\mathbf{Q}_0$ (F.3), then the point in question can be written quite generally as

$$\mathbf{Q}_{\alpha,\gamma} = \begin{bmatrix} Q_1 \\ Q_2 \\ Q_3 \\ Q_4 \\ Q_5 \end{bmatrix} = \begin{bmatrix} Q\cos\alpha \\ 0 \\ Q\sin\alpha\sin 2\gamma \\ Q\sin\alpha\cos 2\gamma \\ 0 \end{bmatrix},\qquad\text{(F.6)}$$

where $0 \le \alpha < \pi/3$ and $0 \le \gamma < \pi$ (if α is outside this region, the unique nearest point is not at $\mathbf{Q}_0$). Now the (θ, ϕ) rotation (F.4) is applied to $\mathbf{Q}_{\alpha,\gamma}$ to generate all the points in $\{Q_i\}$ space following the movement of $\mathbf{Q}_0$ over the sphere, giving

$$\begin{bmatrix} Q_1 \\ Q_2 \\ Q_3 \\ Q_4 \\ Q_5 \end{bmatrix} = D_D \times C_D \times \mathbf{Q}_{\alpha,\gamma},\qquad\text{(F.7)}$$

and this is now the parametrization used in Chapter 3 (3.5).

F.2 ROTATIONS TO DIAGONALIZE $H \otimes h_2$

The matrices $C_D(\theta)$ and $D_D(\phi)$ are just two of three rotation matrices for an Euler angle rotation of a set of $L = 2$ states. The third rotation matrix is

$$B_D(\gamma) = \begin{bmatrix} 1 & 0 & 0 & 0 & 0 \\ 0 & \cos\gamma & 0 & 0 & -\sin\gamma \\ 0 & 0 & \cos 2\gamma & \sin 2\gamma & 0 \\ 0 & 0 & -\sin 2\gamma & \cos 2\gamma & 0 \\ 0 & \sin\gamma & 0 & 0 & \cos\gamma \end{bmatrix}.\qquad\text{(F.8)}$$

In Chapter 5 we discuss the diagonalization of the $H \otimes h_2$ matrix by an orthogonal transformation. The transformation matrix is the product of the three rotation matrices above, together with a matrix in α,

$$A_D(\alpha) = \begin{bmatrix} \cos(\alpha/2) & 0 & 0 & \sin(\alpha/2) & 0 \\ 0 & 1 & 0 & 0 & 0 \\ 0 & 0 & 1 & 0 & 0 \\ -\sin(\alpha/2) & 0 & 0 & \cos(\alpha/2) & 0 \\ 0 & 0 & 0 & 0 & 1 \end{bmatrix}.\qquad\text{(F.9)}$$

If we define

$$\mathcal{T}_D = A_D(\alpha) \times B_D(\gamma) \times C_D(\theta) \times D_D(\phi), \tag{F.10}$$

then the diagonal matrix is given by

$$\mathcal{T}_D \times M_2^H(h) \times \mathcal{T}_D^{-1}. \tag{F.11}$$

F.3 ROTATIONS TO DIAGONALIZE T $\otimes$ h

Listed here are the rotation matrices introduced in Chapter 3 to diagonalize $M^T(h)$:

$$B_P(\gamma) = \begin{bmatrix} \cos\gamma & \sin\gamma & 0 \\ -\sin\gamma & \cos\gamma & 0 \\ 0 & 0 & 1 \end{bmatrix}, \tag{F.12}$$

$$C_P(\theta) = \begin{bmatrix} \cos\theta & 0 & -\sin\theta \\ 0 & 1 & 0 \\ \sin\theta & 0 & \cos\theta \end{bmatrix}, \tag{F.13}$$

and

$$D_P(\phi) = \begin{bmatrix} \cos\phi & \sin\phi & 0 \\ -\sin\phi & \cos\phi & 0 \\ 0 & 0 & 1 \end{bmatrix}. \tag{F.14}$$

The diagonal matrix is given by

$$\mathcal{T}_P \times M^T(h) \times \mathcal{T}_P^{-1}, \tag{F.15}$$

where

$$\mathcal{T}_P = B_P(\gamma) \times C_P(\theta) \times D_P(\phi). \tag{F.16}$$

F.4 REPRESENTATION OF A ROTATING QUADRUPOLE

The components of the $L = 2$ basis for an H representation (E.2) are also the components of a quadrupole, and the parametrization (3.5) can be visualised in terms of a general quadrupole distortion of a sphere. As such, this parametrization was originally proposed for work with the quadrupolar distorted nucleus by Aage Bohr in 1953 (Bohr and Mottelson, 1975). To visualise the parametrization in these terms, note that a general ellipsoidal distortion of a sphere with its principal axes along the Cartesian axes (x, y, z) can be written $Q[\frac{1}{2}(2z^2 - x^2 - y^2)\cos\alpha + \frac{\sqrt{3}}{2}(x^2 - y^2)\sin\alpha]$, or equivalently as the vector of coefficients of the $\{h_i\}$ (E.2):

$$\mathbf{Q}_\alpha = \begin{bmatrix} \cos\alpha \\ 0 \\ 0 \\ \sin\alpha \\ 0 \end{bmatrix}. \tag{F.17}$$

A rotation of this distortion through the Euler angles (γ, θ, ϕ) is then represented by

$$\mathbf{Q} = D_D \times C_D \times B_D \times \mathbf{Q}_\alpha \tag{F.18}$$

and this, which is just the parametrization (3.5) over again, represents a general ellipsoidal distortion in a general orientation with respect to axes fixed in the sphere. The same limitation on the ranges of $\{Q, \alpha, \gamma, \theta, \phi\}$ is needed to ensure that each distinct distortion occurs exactly once. This way of visualizing the parametrization employs the same simplification as Figures 2.5–2.9. The discussion in Section F.1 can be applied to any set of coordinates in a five-dimensional space.

G

Parameters of the Jahn-Teller Minima and Other Stationary Points

TABLE G.1

Parameters Used in Appendix G

(a) Biharmonic parameters for G bases (to normalize, put $g = 1$). (b) Angular parameters for H bases on a spherical subspace (to normalize, put $h = 1$).

$$(a)\begin{cases} g_1 = g \sin\theta \sin\alpha \\ g_2 = g \sin\theta \cos\alpha \\ g_3 = g \cos\theta \sin\beta \\ g_4 = g \cos\theta \cos\beta \end{cases} \qquad (b)\begin{cases} h_1 = h\frac{1}{2}(3\cos^2\theta - 1) \\ h_2 = h\frac{\sqrt{3}}{2} \sin 2\theta \cos\phi \\ h_3 = h\frac{\sqrt{3}}{2} \sin^2\theta \sin 2\phi \\ h_4 = h\frac{\sqrt{3}}{2} \sin^2\theta \cos 2\phi \\ h_5 = h\frac{\sqrt{3}}{2} \sin 2\theta \sin\phi \end{cases}$$

TABLE G.2

Biharmonic Coordinates for the $\{g_i\}$ at the $G \otimes g$, Type IV, Minima

Values of the $\{g_i\}$ at the $G \otimes g$, type IV, minima in terms of the biharmonic coordinates (Table G.1[a])

	g	θ	α	β
min 1	g_0	$\pi/4$	$3\pi/2$	$\pi/2$
min 2	g_0	$\pi/4$	$7\pi/10$	$9\pi/10$
min 3	g_0	$\pi/4$	$3\pi/10$	$\pi/10$
min 4	g_0	$\pi/4$	$11\pi/10$	$17\pi/10$
min 5	g_0	$\pi/4$	$19\pi/10$	$13\pi/10$

$$\text{where} \quad g_0 = \frac{3}{\sqrt{2}} k_g^G \sqrt{\frac{\hbar}{\omega_g}}.$$

TABLE G.3

The $\{g_i\}$ at the $G \otimes g$, Type IV, Minima in Matrix Form

The actual values of the $\{g_i\}$ are the rows of $k_g^G \sqrt{(\hbar/\omega_g)}Q_g^G$. The electronic bases are the rows of $\mathbf{a}_g^G = (\sqrt{2}/3)q_g^G$.

$$Q_g^G = \begin{bmatrix} -3/2 & 0 & 3/2 & 0 \\ p_3 & -p_1 & p_4 & -p_2 \\ p_3 & p_1 & p_4 & p_2 \\ -p_4 & -p_2 & -p_3 & p_1 \\ -p_4 & p_2 & -p_3 & -p_1 \end{bmatrix}, \quad \text{where} \quad \begin{cases} p_1 = (3/2)\cos(3\pi/10) \\ p_2 = (3/2)\cos(\pi/10) \\ p_3 = (3/2)\sin(3\pi/10) \\ p_4 = (3/2)\sin(\pi/10) \end{cases}.$$

TABLE G.4

Normal Coordinates at the D_{3d} Type III Minima of $G \otimes h$

The values of the normal coordinates at the D_{3d}, type III, minima of $G \otimes h$ in terms of the angular parameters in Table G.1(b) on the $\alpha = 0$ trough

	h	θ	ϕ
min 1	h_0	θ_1	$\pi/5$
min 2	h_0	θ_1	$3\pi/5$
min 3	h_0	θ_1	π
min 4	h_0	θ_1	$7\pi/5$
min 5	h_0	θ_1	$9\pi/5$
min 6	h_0	θ_2	$\pi/5$
min 7	h_0	θ_2	$3\pi/5$
min 8	h_0	θ_2	π
min 9	h_0	θ_2	$7\pi/5$
min 10	h_0	θ_2	$9\pi/5$

, where $\begin{cases} h_0 = \sqrt{25/3} \times k_h^G \sqrt{\dfrac{\hbar}{\omega_h}} \\ \cos\theta_1 = \sqrt{(1 + 2/\sqrt{5})/3} \\ \cos\theta_2 = \sqrt{(1 - 2/\sqrt{5})/3} \end{cases}.$

TABLE G.5

The Numerical Values of the $\{h_i\}$ Given in Table G.4

The column norm, the square root of the sum of the squares in each column of this matrix, is $\sqrt{50/3}$. The row norm is $\sqrt{25/3}$. The actual values of the $\{h_i\}$ at a minimum are given by multiplying a row of Q_h^G by $k_h^G \sqrt{\hbar/\omega}$.

$$Q_h^G = \begin{bmatrix} \mu_1 & \mu_2 & \mu_3 & \mu_4 & \mu_5 \\ \mu_1 & -\mu_6 & -\mu_7 & -\mu_6 & \mu_8 \\ \mu_1 & -\mu_9 & 0 & \mu_{10} & 0 \\ \mu_1 & -\mu_6 & \mu_7 & -\mu_6 & -\mu_8 \\ \mu_1 & \mu_2 & -\mu_3 & \mu_4 & -\mu_5 \\ -\mu_1 & \mu_6 & \mu_8 & \mu_6 & \mu_7 \\ -\mu_1 & -\mu_4 & -\mu_5 & -\mu_2 & \mu_3 \\ -\mu_1 & -\mu_{10} & 0 & \mu_9 & 0 \\ -\mu_1 & -\mu_4 & \mu_5 & -\mu_2 & -\mu_3 \\ -\mu_1 & \mu_6 & -\mu_8 & \mu_6 & -\mu_7 \end{bmatrix}, \quad \text{where} \quad \begin{cases} \mu_1 = 1.29099 \\ \mu_2 = 1.95137 \\ \mu_3 = 0.876219 \\ \mu_4 = 0.284701 \\ \mu_5 = 1.41775 \\ \mu_6 = 0.745356 \\ \mu_7 = 0.541533 \\ \mu_8 = 2.29397 \\ \mu_9 = 2.41202 \\ \mu_{10} = 0.921311 \end{cases}.$$

TABLE G.6

Electronic Bases at the D_{3d}, Type III, Minima of $G \otimes h$

The electronic bases at the D_{3d}, type III, minima of $G \otimes h$, given in terms of the normalized biharmonic coordinates (G.1[a]). The angle Ω_D is the dodecahedral angle, the angle between two neighbouring threefold axes.

	θ	α	β	
min 1	Ω_D	$9\pi/10$	$13\pi/10$	
min 2	Ω_D	$17\pi/10$	$9\pi/10$	
min 3	Ω_D	$5\pi/10$	$5\pi/10$	
min 4	Ω_D	$13\pi/10$	$\pi/10$	
min 5	Ω_D	$\pi/10$	$17\pi/10$	, where $\cos\Omega_D = \dfrac{\sqrt{5}-1}{2\sqrt{3}}$.
min 6	$\pi/2 - \Omega_D$	$19\pi/10$	$3\pi/10$	
min 7	$\pi/2 - \Omega_D$	$7\pi/10$	$19\pi/10$	
min 8	$\pi/2 - \Omega_D$	$15\pi/10$	$15\pi/10$	
min 9	$\pi/2 - \Omega_D$	$3\pi/10$	$11\pi/10$	
min 10	$\pi/2 - \Omega_D$	$11\pi/10$	$7\pi/10$	

TABLE G.7

Numerical Values of the Electronic Bases at the D_{3d}, Type III, Minima of $G \otimes h$ in Table G.6

The column norm in this matrix is $\sqrt{5/2}$; the row norm is 1.

$$
\mathbf{a}_h^G = \begin{bmatrix}
v_3 & -v_8 & -v_3 & -v_2 \\
-v_7 & v_6 & v_1 & -v_4 \\
v_9 & 0 & v_5 & 0 \\
-v_7 & -v_6 & v_1 & v_4 \\
v_3 & v_8 & -v_3 & v_2 \\
-v_1 & v_4 & v_7 & v_6 \\
v_3 & -v_2 & -v_3 & v_8 \\
-v_5 & 0 & -v_9 & 0 \\
v_3 & v_2 & -v_3 & -v_8 \\
-v_1 & -v_4 & v_7 & -v_6
\end{bmatrix}, \quad \text{where} \quad
\begin{cases}
v_1 = 0.110264 \\
v_2 = 0.209735 \\
v_3 = 0.288675 \\
v_4 = 0.339358 \\
v_5 = 0.356822 \\
v_6 = 0.549093 \\
v_7 = 0.755761 \\
v_8 = 0.888451 \\
v_9 = 0.934172
\end{cases}.
$$

TABLE G.8

The Electronic Bases at D_{3d}, Type I, Saddle Points of $G \otimes (g \oplus h)$ in Normalized Biharmonic Form (Table G.1[a])

	θ	α	β	
1	$\pi/2 - \Omega_I$	$14\pi/10$	$8\pi/10$	
2	$\pi/2 - \Omega_I$	$12\pi/10$	$14\pi/10$	
3	$\pi/2 - \Omega_I$	0	$10\pi/10$	
4	$\pi/2 - \Omega_I$	$8\pi/10$	$6\pi/10$	
5	$\pi/2 - \Omega_I$	$6\pi/10$	$12\pi/10$	, where $\tan 2\Omega_I = 2$.
6	Ω_I	$14\pi/10$	$18\pi/10$	
7	Ω_I	$12\pi/10$	$4\pi/10$	
8	Ω_I	0	0	
9	Ω_I	$8\pi/10$	$16\pi/10$	
10	Ω_I	$6\pi/10$	$2\pi/10$	

TABLE G.9

The Values of the $\{g_i\}$ at D_{3d}, Type I, Saddle Points of $G \otimes (g \oplus h)$

The values are given in terms of biharmonic coordinates. The $\{h_i\}$ at this D_{3d} stationary

point are given by the array in Table G.4, but with $h_0 = \sqrt{3} \times k_h^G \sqrt{\frac{\hbar}{\omega_h}}$.

	g	θ	α	β
1	g_0	θ_3	$9\pi/10$	$13\pi/10$
2	g_0	θ_3	$17\pi/10$	$9\pi/10$
3	g_0	θ_3	$5\pi/10$	$5\pi/10$
4	g_0	θ_3	$13\pi/10$	$\pi/10$
5	g_0	θ_3	$\pi/10$	$17\pi/10$
6	g_0	$\pi/2 - \theta_3$	$19\pi/10$	$3\pi/10$
7	g_0	$\pi/2 - \theta_3$	$7\pi/10$	$19\pi/10$
8	g_0	$\pi/2 - \theta_3$	$15\pi/10$	$15\pi/10$
9	g_0	$\pi/2 - \theta_3$	$3\pi/10$	$11\pi/10$
10	g_0	$\pi/2 - \theta_3$	$11\pi/10$	$7\pi/10$

$$, \quad \text{where} \quad \begin{cases} g_0 = \sqrt{3} \times k_g^G \sqrt{\frac{\hbar}{\omega_g}} \\ \cos\theta_3 = 0.934172 \end{cases}.$$

TABLE G.10

The Electronic Bases at the D_{2h}, Type II, Saddle-Points of

$G \otimes (g \oplus h)$ in Biharmonic Form

	θ	α	β
1	$\pi/4 + \Omega_I$	0	0
2	$\pi/4 + \Omega_I$	$18\pi/10$	$6\pi/10$
3	$\pi/4 + \Omega_I$	$12\pi/10$	$4\pi/10$
4	$\pi/4 + \Omega_I$	$16\pi/10$	$12\pi/10$
5	$\pi/4 + \Omega_I$	$14\pi/10$	$18\pi/10$
6	$\pi/4 - \Omega_I$	$10\pi/10$	0
7	$\pi/4 - \Omega_I$	$8\pi/10$	$6\pi/10$
8	$\pi/4 - \Omega_I$	$2\pi/10$	$4\pi/10$
9	$\pi/4 - \Omega_I$	$16\pi/10$	$2\pi/10$
10	$\pi/4 - \Omega_I$	$14\pi/10$	$8\pi/10$
11	$\pi/4$	$5\pi/10$	$5\pi/10$
12	$\pi/4$	$13\pi/10$	$1\pi/10$
13	$\pi/4$	$17\pi/10$	$9\pi/10$
14	$\pi/4$	$11\pi/10$	$7\pi/10$
15	$\pi/4$	$19\pi/10$	$3\pi/10$

$$, \quad \text{where} \quad \tan 2\Omega_I = 2.$$

TABLE G.11

The Values of the $\{h_i\}$ at the D_{2h}, Type II, Saddle Points of $G \otimes (g \oplus h)$
They are given here explicitly, rather than in parametric form, because they do not occur exactly on an $\alpha = 0$ surface. They are the numbers in the table multiplied by $k_h^G \sqrt{\frac{\hbar}{\omega_h}}$.

	h_1	h_2	h_3	h_4	h_5	
1	λ_{10}	λ_{16}	0	λ_5	0	$\lambda_1 = 0.105573$
2	λ_{10}	λ_7	λ_2	$-\lambda_3$	λ_{15}	$\lambda_2 = 0.200811$
3	λ_{10}	λ_7	$-\lambda_2$	$-\lambda_3$	$-\lambda_{15}$	$\lambda_3 = 0.276393$
4	λ_{10}	$-\lambda_{12}$	$-\lambda_4$	λ_1	λ_9	$\lambda_4 = 0.32492$
5	λ_{10}	$-\lambda_{12}$	λ_4	λ_1	$-\lambda_9$	$\lambda_5 = 0.341641$
6	$-\lambda_{10}$	$-\lambda_5$	0	$-\lambda_{16}$	0	$\lambda_6 = 0.618034$
7	$-\lambda_{10}$	$-\lambda_1$	$-\lambda_9$	λ_{12}	$-\lambda_4$	$\lambda_7 = 0.723607$
8	$-\lambda_{10}$	$-\lambda_1$	λ_9	λ_{12}	λ_4	$\lambda_8 = 1.17557$
9	$-\lambda_{10}$	λ_3	λ_{15}	$-\lambda_7$	$-\lambda_2$	$\lambda_9 = 1.37638$
10	$-\lambda_{10}$	λ_3	$-\lambda_{15}$	$-\lambda_7$	λ_2	$\lambda_{10} = 1.54919$
11	0	$-\lambda_{14}$	0	λ_{14}	0	$\lambda_{11} = 1.61803$
12	0	$-\lambda_6$	λ_8	$-\lambda_{11}$	$-\lambda_{13}$	$\lambda_{12} = 1.89443$
13	0	$-\lambda_6$	$-\lambda_8$	$-\lambda_{11}$	λ_{13}	$\lambda_{13} = 1.90211$
14	0	λ_{11}	$-\lambda_{13}$	λ_6	$-\lambda_8$	$\lambda_{14} = 2.$
15	0	λ_{11}	λ_{13}	λ_6	λ_8	$\lambda_{15} = 2.22703$

where $\lambda_{16} = 2.34164$

TABLE G.12

The $\{g_i\}$ at the Type II Saddle Points of $G \otimes (g \oplus h)$

The electronic bases that correspond to D_{2h} saddle points in $G \otimes (g \oplus h)$ in general are threefold degeneracies if the g modes alone are coupled, and the symmetry of the distortion is T_d. There are thus only five sets of the $\{g_i\}$, and these are given in biharmonic form in the following table:

	g	θ	α	β
1	g_0	$\pi/4$	$13\pi/10$	$11\pi/10$
2	g_0	$\pi/4$	$17\pi/10$	$19\pi/10$
3	g_0	$\pi/4$	$\pi/10$	$7\pi/10$
4	g_0	$\pi/4$	$9\pi/10$	$3\pi/10$
5	g_0	$\pi/4$	$5\pi/10$	$15\pi/10$

, where $g_0 = \frac{1}{\sqrt{2}} \times k_g^G \sqrt{\frac{\hbar}{\omega_g}}$.

TABLE G.13

Bases and $\{g_i\}$ at the D_{3d} Minima of $H \otimes g$

The values of the $\{g_i\}$ at the minima are given by Q_g^H and the bases at the minima are given by $\mathbf{a}_g^H$. These two equations illustrate the duality of $G \otimes h$ and $H \otimes g$.

$$Q_g^H = \sqrt{64/3}\, \mathbf{a}_h^G \quad (\mathbf{a}_h^G \text{ from Table G.7})$$

$$\mathbf{a}_g^H = \sqrt{3/25}\, Q_h^G \quad (Q_h^G \text{ from Table G.5})$$

TABLE G.14

Bases and $\{h_i^4\}$ at the D_{5d} Minima of $H \otimes h_4$

The mimima occur at the vertices of an icosahedron on the $\alpha = 0$ surface in h space; these points are at

	θ	ϕ
1	0	any
2	θ_I	0
3	θ_I	$2\pi/5$
4	θ_I	$4\pi/5$
5	θ_I	$6\pi/5$
6	θ_I	$8\pi/5$

$,$ where $\cos\theta_I = 1/\sqrt{5}$.

They translate into the five sets of values of h_i by means of the transformation (E.2) to give

	h_1	h_2	h_3	h_4	h_5
1	h_0	0	0	0	0
2	$-h_0/5$	η_1	0	η_1	0
3	$-h_0/5$	η_2	η_4	η_3	η_5
4	$-h_0/5$	η_3	$-\eta_5$	η_2	η_4
5	$-h_0/5$	η_3	η_5	η_2	$-\eta_4$
6	$-h_0/5$	η_2	$-\eta_4$	η_3	$-\eta_5$

where

$$
\begin{cases}
h_0 = \sqrt{108}k\sqrt{\hbar/\omega} \\
\eta_1 = (2\sqrt{3}h_0/5) \\
\eta_2 = (2\sqrt{3}h_0/5)\cos 2\pi/5 \\
\eta_3 = (2\sqrt{3}h_0/5)\cos 4\pi/5 \\
\eta_4 = (2\sqrt{3}h_0/5)\sin 4\pi/5 \\
\eta_5 = (2\sqrt{3}h_0/5)\sin 2\pi/5
\end{cases}
$$

.

The normalized bases at these points are given by the same array, taken with $h_0 = 1$.

H

Cited References and Bibliography

H.1 CITED REFERENCES

Abragam, A., and Bleaney, B. 1970. *Electron Paramagnetic Resonance of Transition Ions*. Oxford: Oxford University Press.

Aitchison, I. J. R. 1988. *Physica Scripta* **T23**, 12–20.

Anderson, F. G., Ham, F. S., and Grossman, G. 1984. *Materials Sci. Forum* **83-87**, 475–80.

Antropov, V. P., Gunnarson, A., and Liechtenstein, A. I. 1993. *Phys. Rev. B* **48**, 7651–64.

Apsel, S. E., Chancey, C. C., and O'Brien, M. C. M. 1992. *Phys. Rev. B* **45**, 5251–61.

Auerbach, A., and Kivelson, S. 1985. *Nucl. Phys.* **B257**, 799–858.

Auerbach, A., Manini, N., and Tosatti, E. 1994. *Phys. Rev. B* **49**, 12998–3007.

Ballhausen, C. J. 1965. *Theor. Chim. Acta* (Berlin) **3**, 368–74.

Barut, A. O., and Rączka, R. 1986. *Theory of Group Representations and Applications*. Singapore: World Scientific Publishing Company.

Bendale, R. D., Stanton, J. F., and Zerner, M. C. 1992. *Chem. Phys. Lett.* **194**, 467–71.

Berry, M. V. 1984. *Proc. R. Soc. London Ser. A* **392**, 45–57.

Bersuker, I. B., and Polinger, V. Z. 1989. *Vibronic Interactions in Molecules and Crystals*. Berlin: Springer-Verlag.

Bleaney, B., and Bowers, K. D. 1952. *Proc. Phys. Soc.* (London) **A65**, 667–8.

Bohr, B., and Mottelson, B. R. 1975. *Nuclear Structure* **2**, 677–92.

Born, M., and Huang, K. 1954. *Dynamical Theory of Crystal Lattices*. Oxford: Oxford University Press.

Born, M., and Oppenheimer, R. 1927. *Ann. Phys.* **84**, 457–84.

Boyd, P. D. W., Bhyrappa, P., Paul, P., Stinchcombe, J., Bolskar, R. D., Sun, Y., and Reed C. A. 1995. *J. Am. Chem. Soc.* **117**, 2907–14.

Ceulemans, A. 1994. *Topics in Current Chemistry* **171** 27–67.

Ceulemans, A., and Fowler, P. W. 1989. *Phys. Rev. A* **39**, 481–93.

Ceulemans, A., and Fowler, P. W. 1990. *J. Chem Phys.* **93**, 1221–34.

Ceulemans, A., Fowler, P. W., and Vos, I. 1993. *J. Chem. Phys.* **100**, 5491–500

Ceulemans, A., and Vanquickenbourne, L. G. 1989. *Structure and Bonding* **71** 125–59.

Chancey, C. C., and O'Brien, M. C. M. 1988. *J. Phys. A* **21**, 3347–53.

Chancey, C. C., and O'Brien, M. C. M. 1989. *J. Phys.: Condens. Matter* **1**, 47–68.

Cho, K. 1968. *J. Phys. Soc. Japan* **25**, 1372–87.

Chung, F., and Sternberg, S. 1995. *American Scientist* **81**, 56–71.

Coffman, R. E. 1965. *Phys. Letters* **19**, 475–6.

Cohen, M. L. 1993. *Materials Sci. and Engin.* **B19**, 111–6.

Coxeter, H. M. S. 1971. "Virus macromolecules and geodesic domes." In *A Spectrum of Mathematics*, edited by J. C. Butcher. Oxford: Oxford University Press.

Cullerne, J. P. 1995. "The Jahn-Teller Effect in Icosahedral Symmetry." Thesis, Oxford University.

Cullerne, J. P., Angelova, M. N., and O'Brien, M. C. M. 1995. *J. Phys.: Condens. Matter* **7**, 3247–69.

Cullerne, J. P., and O'Brien, M. C. M. 1994. *J. Phys.: Condens. Matter* **6** 9017–41.

Danieli, R., Denisov, V. N., Ruani, G., Zamboni, R., Taliani, Zakhidov, A. A., Ugawa, A., Imaeda, K., Yakushi, K., Inokuchi, H., Kikuchi, K., Ikemoto, I., Suzuki, S., and Achiba, Y. 1992. *Solid State Commun.* **81**, 257–60.

de Coulon, V., Martins, J. L., and Reuse, F. 1992. *Phys. Rev. B* **45**, 13671–5.

Dresselhaus, G., Dresselhaus, M. S., and Eklund, P. C. 1992. *Phys. Rev.* **45**, 6923–30.

Dresselhaus, M. S., Dresselhaus, G., and Saito, R. 1993. *Materials Sci. and Engin.* **B19**, 122–8.

Dunn, J. L., and Bates, C. A. 1995. *Phys. Rev. B* **52** 5996–6005.

Echt, O., Sattler, K., and Rechnagel, E. 1981. *Phys. Rev. Lett.* **47**, 1121–4.

Edmonds, A. R. 1960. *Angular Momentum in Quantum Mechanics*, 2nd ed. Princeton: Princeton University Press.

Eisenberg, J. M., and Greiner, W. 1970. *Nuclear Theory*. Vol. 1, *Nuclear Models*. Amsterdam: North-Holland.

Englman, R. 1972. *The Jahn-Teller Effect in Molecules and Crystals*. London: Wiley-Interscience.

Essén, H. 1977. *Intl. J. Quantum Chem.* **12**, 721–735.

Evangelou, S. N., O'Brien, M. C. M., and Perkins, R. S. 1980. *J. Phys. C* **13**, 4175–98.

Fagerström, J., and Stafström, S. 1993. *Phys. Rev. B* **48**, 11367—74.

Fischer, G. 1984. *Vibronic Coupling*. London: Academic Press.

———. 1989. "Vibronic Coupling, Bases." In *Vibronic Processes in Inorganic Chemistry*, edited by Colin D. Flint. Dordrecht: Kluwer Academic Publishers.

Fletcher, J. R., O'Brien, M. C. M., and Evangelou, S. N. 1979. *J. Phys. A* **13** 2035–47.

Fowler, P. W., and Manolopoulos, D. E. 1995. *An Atlas of Fullerenes.* Oxford: Oxford University Press.

Fowler, P. W., Cremona, J. E., and Steer, J. I. 1988. *Theor. Chim. Acta* **73**, 1–26.

Fowler, P. W., and Woolrich, J. 1986. *Chem. Phys. Lett..* **127**, 78–83.

Fulara, J., Jakobi, M., and Maier, J. P. 1993. *Chem. Phys. Lett.* **211**, 227–34.

Gasyna, Z., Andrews, L., and Schatz, P. N. 1992. *J. Phys. Chem.* **96**, 1525–7.

Golding, R. M. 1973. *Mol. Phys.* **26**, 661–72.

Groenen, E. J. J., Poluektov, O. G., Matsushita, M., Schmidt, J., and van der Waals, J. H. 1992. *Chem. Phys. Lett.* **197**, 314–8.

Gu, B., Li, Z., and Zhu, J. 1993. *J. Phys.: Condens. Matter* **5**, 5255–60.

Haddon, R. C., Brus, L. E., and Raghavachari, K. 1986. *Chem. Phys. Lett.* **125**, 459–64.

Ham, F. S. 1965. *Phys. Rev.* **138**, 727–39.

———. 1968. *Phys. Rev.* **166**, 307–21.

———. 1987. *Phys. Rev. Lett.* **58**, 725–8.

———. 1990. *J. Phys.: Cond. Matter* **2**, 1163–77

Hawkins, J. M., Meyer, A., Lewis, T. A., Loren, S., and Hollander, F. J. 1991. *Science* **252**, 312–3.

Head, J., and Zerner, M. C. 1985. *Chem. Phys. Lett.* **122**, 264–70.

Heath, G. A., McGrady, J. E., and Martin, R. 1992. *J. Chem. Soc., Chem. Commun.* **17**, 1272–74.

Henry, C. H., Schnatterly, S. E., and Schlichter, C. P. 1965. *Phys. Rev.* **137**, A583–602.

Herzberg, G. 1966. *Molecular Structure III Electronic Spectra and Electronic Structure of Polyatomic Molecules.* Princeton, NJ: Van Nostrand.

Huang, Z. H., Feuchtwang, T. E., Cutler, P. H., and Kazes, E. 1990. *Phys. Rev. A* **41**, 32–41.

Jahn, H. A. 1938. *Proc. R. Soc. London Ser. A* **164**, 117–31.

Jahn, H. A., and Teller, E. 1936. Abstract in *Phys. Rev.* **49**, 874.

Jahn, H. A., and Teller, E. 1937. *Proc. Roy. London Soc. Ser. A* **161**, 220–35.

Jishi, R. A., and Dresselhaus, M. S. 1992. *Phys. Rev. B* **45**, 6914–18.

Judd, B. R. 1957. *Proc. R. Soc. London Ser. A* **241**, 122–31.

———. 1963. *Operator Techniques in Atomic Spectroscopy.* New York: McGraw-Hill.

———. 1974. *Can. J. Phys.* **52**, 999–1044.

———. 1984. *Advances in Chemical Physics* **57**, 247–309.

Kapur, P. L., and Peierls, R. 1937. *Proc. Roy. London Soc. A* **163**, 606–10.

Kato, T., Kodama, T., Shida, T., Nakagawa, T., Matsui, Y., Suzuki, S., Shiromaru, H., Yamauchi, K., and Achiba, Y. 1991. *Chem. Phys. Lett.* **180**, 446–50.

Khaled, M. M., Carlin, R. T., Trulove, P. C., Eaton, G. R., and Eaton, S. S. 1994. *J. Am. Chem. Soc.* **116**, 3465–74.

Khlopin, V. P., Polinger, V. Z., and Bersuker, I. B. 1978. *Theor. Chim. Acta* (Berlin) **48**, 87-101.

Kimura K. 1993. *Materials Sci. and Engin.* **B19** 67-71.

Kimura, K., Hori, A., Yamashita H. and Ino H. 1993. *Phase Transitions* **44** 173–82.

Koga, N., and Morokuma, K. 1992. *Chem. Phys. Lett.* **196**, 191–5.

Kondo, H., Momose, T., and Shida, T. 1995. *Chem. Phys. Lett.* **237**, 111–114.

Kronig, R. de L. 1930. *Band Spectra and Molecular Structure.* Cambridge: Cambridge University Press.

Kroto, H. W., Heath, J. R., O'Brien, S. C., Curl, R. F., and Smalley, R. E. 1985. *Nature* **318**, 162–3.

Lane, P. A., Swanson, L. S., Ni, Q-X, Shinar, J., Engel, J. P., Barton, T. J. and Jones, L. 1992. *Phys. Rev. Lett.* **68**, 887–90.

Lannoo, M., Baraff, G. A., Schlüter, M., and Tomanek, D. 1991. *Phys. Rev. B* **44**, 12106–9.

Longuet-Higgins, H. C. 1961. *Advan. Spectry.* **2**, 429–72.

Longuet-Higgins, H. C., Öpik, U., Pryce, M. H. L., and Sack, H. 1958. *Proc. R. Soc. London Ser. A* **244**, 1–16.

Louck, J. D. 1960. *J. Mol. Spect.* **4**, 298–333.

Manini, N., Tosatti, E., and Auerbach, A. 1994. *Phys. Rev. B* **49**, 13008–16.

McLachlin, A. D. 1961. *Mol. Phys.* **4**, 417–23.

Negri, F., Orlandi, G., and Zerbetto, F. 1988. *Chem. Phys. Lett.* **144**, 31–37.

———. 1992. *J. Am. Chem. Soc.* **114**, 2909–13.

O'Brien, M. C. M. 1969. *Phys. Rev.* **187**, 407–18.

———. 1971. *J. Phys. C: Solid State Physics* **4**, 2524–36.

———. 1972. *J. Phys. C: Solid State Physics* **5**, 2045–63.

———. 1976. *J. Phys. C: Solid State Physics* **9**, 3153–64.

———. 1989. *J. Phys. A: Math. Gen.* **22**, 1779–97.

———. 1996. *Phys. Rev. B* **53**, 3775–89.

O'Brien, M. C. M., and Chancey, C. C. 1993. *Am. J. Phys.* **61**, 688–97.

Öpik, U., and Pryce, M. H. L. 1957 *Proc. Roy. Soc. London Ser. A* **238** 425–47.

Parks, E. K., Winter, B. J., Klots, T. D., and Riley, S. J. 1991. *J. Chem. Phys.* **94**, 1882–1902.

Parlett, B. N. and Reid, J. P. 1980. *AERE Harwell Report* CSS 83

Paul, P., Xie, Z., Bau, R., Boyd, P. D. W., and Reed, C. A. 1994. *J. Am. Chem. Soc.* **116**, 4145–6.

Pennington, C. H., and Stenger, V. A. 1996. *Rev. Mod. Phys.* **68**, 855–910.

Pickett, W. E. 1994. "Electrons and Phonons in C_{60}-Based Materials." In *Solid State Physics*, Vol. 48, edited by H. Ehrenreich and F. Spaepen. San Diego: Academic Press. pp. 225–347.

———. 1991. *Nature* **351**, 602–3.

Polinger, V. Z. 1974. *Fiz. Tverd. Tela (Sov. Phys.-Solid State)* **16**, 2578–83.

Pooler, D. R. 1978. *J. Phys. A: Math. Gen.* **11**, 1045–55.

————. 1980. *J. Phys. C: Solid Phys.* **13**, 1029–42.

————. 1984. "Numerical Diagonalization Techniques in the Jahn-Teller Effect." In *The Dynamical Jahn-Teller Effect in Localized Systems*, edited by Y. E. Perlin and M. Wagner. Amsterdam: Elsevier, 199–250.

Pooler, D. R., and O'Brien, M. C. M. 1977. *J. Phys. C: Solid State Physics* **10**, 3769–91.

Ragnfagni, A. 1977. *Phys. Lett.* **62A**, 395–6.

Regev, A., Gamliel, D., Meiklyar, V., Michaeli, S., and Levanon, H. 1993. *J. Phys. Chem.* **97**, 3671–9.

Renner, R. 1934. *Z. Physik* **92**, 172–93.

Rose, J., Smith, D., Williamson, B. E., Schatz, P. N., and O'Brien, M. C. M. 1986. *J. Phys. Chem.* **90**, 2608–15.

Röthlisberger U., Andreoni, W., and Giannozzi, P. 1992. *J. Chem. Phys.* **96**, 1248–56.

Schlüter, M., Lannoo, M., Needels, M., Baraff, G. A., and Tománek, D. 1992. *Phys. Rev. Lett.* **68**, 526–9. Also published in *J. Phys. Chem. Solids* (1992) **53**, 1473.

————. 1993. *Materials Sci. and Engin.* **B19**, 129–134.

Sloncjewski, J. C. 1963. *Phys. Rev.* **131**, 1596–610.

Stinchcombe, J., Pénicaud, A., Bhyrappa, P., Boyd, P. D. W., and Reed, C. A. 1993. *J. Am. Chem. Soc.* **115**, 5212–7.

Sturge, M. D. 1967. "The Jahn-Teller Effect in Solids." In *Solid State Physics*, Vol. 20, edited by F. Seitz, D. Turnbull, and H. Ehrenreich. New York: Academic Press.

Sugano, S., Tanabe, Y., and Kamimura, H. 1970. "Multiplets of Transition-Metal Ions in Crystals." *Pure and Applied Physics,* vol. **33**, Academic Press.

Surján, P. R., Németh K., Bennati, M., Grupp, A., and Mehring, M. 1996. *Chem. Phys. Lett.* **251**, 116–8.

Surján, P. R., Udvardi, L., and Németh, K. 1994. *J. Molecular Struct.* **311**, 55–68.

Takahashi, T. 1993. *Materials Sci. and Engin.* **B19**, 117–121.

Teller, E. 1982. *Physica A* **114**, 14–18.

————. 1972. "An historical note" on a page inserted before the preface. In *The Jahn-Teller Effect in Molecules and Crystals*, by R. Englman. London: Wiley-Interscience.

Tinkham, M. 1964. *Group Theory and Quantum Mechanics*. New York: McGraw-Hill.

Trulove, P. C., Carlin, R. T., Eaton, G. R., and Eaton, S. S. 1995. *J. Am. Chem. Soc.* **117**, 6265–72.

Tycko, R., Dabbagh, G., Fleming, R. M., Haddon, R. C., Makhija, A. V., and Zahurak, S. M. 1991. *Phys. Rev. Lett.* **67**, 1886–9.

Varma, C. M., Zaanen, J., and Raghavachari, K. 1991. *Science* **254**, 989–92.

Wang, C-L., Wang, W-Z., Liu, Y-L., Su, Z-B., and Yu, L. 1994. *Phys. Rev. B* **50**, 5676–79.

Wasielski, M. R., O'Neil, M. P., Lykke, K. R., Pellin, M. J., and Gruen, D. M. 1991. *J. Am. Chem. Soc.* **113**, 2774–5.

Watkins, G. D. 1986. "The Lattice Vacancy in Silicon." In *Deep Centers in Semi-Conductors*, edited by S. T. Pantelides, 147–183.

Wilczek, F., and Zee, A. 1984. *Phys. Rev. Lett.* **52**, 2111–4.

Yannoni, C. S., Bernier, P. P., Bethune, D. S., Meijer, G., and Salen, J. R. 1991. *J. Am. Chem. Soc.* **113**, 3190–2.

Zhang, B. L., Wang, C. Z., Ho, K. M., Xu, C. H., and Chan, C. T. 1992. *J. Chem. Phys.* **97**, 5007–11.

Zhang, Q.-M., Yi, J.-Y., and Bernholc, J. 1991. *Phys. Rev. Lett.* **66**, 2633–6.

Zhou, P., Wang, K. A., Wang, Y., Ecklund, P. C., Dresselhaus, M. S., Dresselhaus, G., and Jishi, R. A. 1992. *Phys. Rev. B* **46**, 2595–605.

Zwanziger, J. W., Koenig, M., and Pines, A. 1990. *Annu. Rev. Phys. Chem.* **41**, 601–46.

H.2 BIBLIOGRAPHY

The references listed below are collected by topic to form a source of further information. Some have been cited in the text, others not.

H.2.1 Molecular Quantum Mechanics

Born, M., and Huang, K. 1954. *Dynamical Theory of Crystal Lattices.* Oxford: Oxford University Press.

Flint, C. D., ed. 1989. *Vibronic Processes in Inorganic Chemistry.* Dordrecht: Kluwer Academic Publishers. (A collection of papers from a NATO Advanced Science Institute Series on vibronic and Jahn-Teller coupling in molecules and crystals.)

McWeeny, R. 1992. *Methods of Molecular Quantum Mechanics.* 2nd ed. London: Academic Press. (An excellent general introduction that provides background to many of the techniques used in this book.)

H.2.2 Group Theory and Techniques

Joshi, A. W. 1982. *Elements of Group Theory for Physicists.* 3rd ed. New York: John Wiley & Sons.

Judd, B. R, 1963. *Operator Techniques in Atomic Spectroscopy.* New York: McGraw-Hill

Rotenberg, M., Bivins, R., Metropolis, N., and Wooten, J. K. 1959. *The 3-j and 6-j Symbols.* MIT, The Technology Press.

Tinkham, M. 1964. *Group Theory and Quantum Mechanics*. New York: McGraw-Hill.

Wybourne, B. G. 1970. *Symmetry Principles and Atomic Spectroscopy*. Including an appendix of tables by P. H. Butler. New York: Wiley-Interscience.

Wybourne, B. G. 1974. *Classical Groups for Physicists*. New York: Wiley-Interscience.

H.2.3 The Icosahedral Group

Herzberg, G. 1966. *Molecular Structure III Electronic Spectra and Electronic Structure of Polyatomic Molecules*. Princeton, NJ: Van Nostrand. Contains character tables and tables of products of irreducible representations for both the single and double groups.

Golding, R. M. 1973. "Symmetry Coupling Coefficients for the Icosahedral Double Group." In *Molecular Physics* **26**, 661–72.

Judd, B. R. 1957. "A Crystal Field of Icosahedral Symmetry." In *Proc. Roy. Soc.* **A241**, 122–31. (Gives $I \to SO[3]$ reduction.)

H.2.4 The Jahn-Teller Effect

Bersuker, I. B., and Polinger, V. Z. 1989. *Vibronic Interactions in Molecules and Crystals*. Berlin: Springer-Verlag.

Ceulemans, A., and Vanquickenbourne, L. G. 1989. "The Epikernal Principle." *Structure and Bonding* **71**, 125–59.

Englman, R. 1972. *The Jahn-Teller Effect in Molecules and Crystals*. London: Wiley-Interscience.

Ham, F. S. 1972. "Jahn-Teller Effect in Electron Paramagnetic Resonance Spectra." In *Electron Paramagnetic Resonance*, edited by S. Gershwin. New York: Plenum.

Judd, B. R. 1984. "Group theoretical approaches." *The Dynamical Jahn-Teller Effect in Localized Systems*, edited by Y. E. Perlin and M. Wagner. Amsterdam: Elsevier Science Publishers.

O'Brien, M. C. M. 1981. "Vibronic Spectra with Jahn-Teller interactions." *Vibrational Spectra and Structure*, edited by J. R. Durig. Elsevier **10**, 321–94.

Sturge, M. D. 1967. "The Jahn-Teller Effect in Solids." In *Solid State Physics*, Vol. 20. edited by F. Seitz, D. Turnbull, and H. Ehrenreich. New York: Academic Press.

H.2.5 Icosahedral Systems

Ceulemans, A., and Fowler, P. W. 1989. "SO(4) Symmetry and the Static Jahn-Teller Effect in Icosahedral Molecules." *Phys. Rev. A* **39**, 481–93.

Ceulemans, A., and Fowler, P. W. 1990. "The Jahn-Teller Instability of Fivefold Degenerate States in Icosahedral Molecules." *J. Chem. Phys.* **93**, 1221–34.

Cullerne, J. P., Angelova, M. N., and O'Brien, M. C. M. 1995. "The Jahn-Teller Effect in Icosahedral Symmetry: Extension of Ham Factors in Strongly Coupled Systems." *J. Phys.: Condens. Matter* **7**, 3247–69.

Cullerne, J. P., and O'Brien, M. C. M. 1994. "The Jahn-Teller Effect in Icosahedral Symmetry: Ground-State Topography and Phases." *J. Phys.: Condens. Matter* **6**, 9017–41.

Dresselhaus, M. S., Dresselhaus, G., and Saito, S. 1993. "Group Theoretical Concepts for C_{60} and Other Fullerenes." *Materials Sci. and Engin.* **B19**, 122–8.

Fowler, P. W., and Ceulemans, A. 1993. "Spin-Orbit Coupling Coefficients for Icosahedral Molecules." *Theoretica Chemica Acta.* **86**, 315–42.

Khlopin, V. P., Polinger, V. Z., and Bersuker, I. B. 1978. "The Jahn-Teller Effect in Icosahedral Molecules and Complexes." *Theor. Chim. Acta (Berlin)* **48**, 87–101.

Pooler, D. R. 1980. "Continuous Group Invariences of Linear Jahn-Teller Systems: II. Extension and Application to Icosahedral Systems." *J. Phys. C: Solid State Phys.* **13**, 1029–42.

H.2.5.1. C_{60}

Dresselhaus, M. S., Dresselhaus, G., and Eklund, P. C. 1996. *Science of Fullerenes and Carbon Nanotubules.* San Diego: Academic Press.

Koga, N., and Morokuma, K. 1992. "Ab Initio MO Study of the C_{60} Anion Radical: The Jahn-Teller Distortion and Electronic Structure." *Chem. Phys. Lett.* **196**, 191–5.

Paul, P., Xie, Z., Bau, R., Boyd, P. D. W., and Reed, C. A. 1994. "Ordered Structure of a Distorted C_{60}^{2-} Fulleride Ion." *J. Am. Chem. Soc.* **116**, 4145–6.

Surján, P. R., Uvardi, L., and Németh, K. 1994. "Electronic Excitations in Fullerenes: Jahn-Teller Distorted Structures of C_{60}." *J. Molecular Struct.* **311**, 55–68.

H.2.6 The Berry Phase

Aitchison, I. J. R. 1988. "Berry Topological Phase in Quantum-Mechanics and Quantum-Field Theory." *Physica Scripta* **T23**, 12–20.

Berry, M. V. 1984. "Quantal Phase-Factors Accompanying Adiabatic Changes." *Proc. Roy. Soc. London Ser. A* **392**, 45–57.

H.2.7 Spin Resonance

Abragam, A., and Bleaney, B. 1970. *Electron Paramagnetic Resonance of Transition Ions.* Oxford: Oxford University Press.

H.2.7.1. $^3C_{60}{}^*$

Groenen, E. J. J., Poluektov, O. G., Matsushita, M., Schmidt, J., and van der Waals, J. H. 1992. "Triplet Excitation of C_{60} and the Structure of the Crystal at 1.2 K." *Chem. Phys. Lett.* **197**, 314–8.

Lane, P. A., Swanson, L. S., Ni, Q-X, Shinar, J., Engel, J. P., Barton, T. J., and Jones, L. 1992. "Dynamics of Photoexcited States in C_{60}: An Optically Detected Magnetic Resonance, ESR, and Light-Induced ESR Study." *Phys. Rev. Lett.* **68**, 887–90.

Regev, A., Gamliel, D., Meiklyar, V., Michaeli, S., and Levanon, H. 1993. "Dynamics of $^3C_{60}$ Probed by Electron Paramagnetic Resonance. Motional Analysis in Isotropic and Liquid Crystalline Matrices." *J. Phys. Chem.* **97**, 3671–9.

Wasielski, M. R., O'Neil, M. P., Lykke, K. R., Pellin, M. J., and Gruen, D. M. 1991. "Triplet States of Fullerenes C_{60} and C_{70}: Electron Paramagnetic Resonance Spectra, Photophysics, and Electronic Structures." *J. Am. Chem. Soc.* **113**, 2774–5.

Wei, X., and Vardeny, Z. V. 1995. "Spin Dynamics of Triplet Photoexcitations in C_{60}: Evidence for a Dynamic Jahn-Teller Effect." *Phys. Rev. B* **52** R2317–20.

H.2.7.2. $C_{60}{}^{n-}$

Boyd, P. D. W., Bhyrappa, P., Paul, P., Stinchcombe, J., Bolskar, R. D., Sun, Y., and Reed, C. A. 1995. "The $C_{60}{}^{2-}$ Fulleride Ion." *J. Am. Chem. Soc.* **117**, 2907–14.

Khaled, M. M., Carlin, R. T., Trulove, P. C., Eaton, G. R., and Eaton, S. S. 1994. "Electrochemical Generation and Electron Paramagnetic Resonance Studies of $C_{60}{}^-$, $C_{60}{}^{2-}$ and $C_{60}{}^{3-}$." *J. Am. Chem. Soc.* **116**, 3465–74.

Stinchcombe, J., Pénicaud, A., Bhyrappa, P., Boyd, P. D. W., and Reed C. A. 1993. "Buckminsterfulleride(1-) Salts: Synthesis, EPR, and the Jahn-Teller Distortion of $C_{60}{}^-$." *J. Am. Chem. Soc.* **115**, 5212–7.

Trulove, P. C., Carlin, R. T., Eaton, G. R., and Eaton, S. S. 1995. "Determination of the Singlet-Triplet Energy Separation for $C_{60}{}^{2-}$ in DMSO by Electron Paramagnetic Resonance." *J. Am. Chem. Soc.* **117**, 6265–72.

H.2.8 Numerical Methods

Pooler, D. R. 1984. "Numerical Diagonalization Techniques in the Jahn-Teller Effect." *The Dynamical Jahn-Teller Effect in Localized Systems*, edited by Y. E. Perlin and M. Wagner. Elsevier, 199–250.

H.2.9 Superconductivity in the Fullerides

Cohen, M. L. 1993. "Theory of Normal and Superconducting Properties of Fullerene-Based Solids." *Materials Sci. and Engin.* **B19**, 111–16.
Pennington, C. H., and Stenger, V. A. "Nuclear Magnetic Resonance of C_{60} and Fulleride Superconductors." *Phys. Rev. Mod.* **68**, 855–910.
Schlüter, M., Lannoo, M., Needels, M., Baraff, G. A., and Tománek, D. 1992. "Superconductivity in Alkali Intercalated C_{60}." *J. Phys. Chem. Solids* **53**, 1473–85.
———. 1993. "Superconductivity in Alkali-Intercalated C_{60}." *Materials Sci. and Engin.* **B19**, 129–134.

H.2.10 Molecular Spectra

Fulara, J., Jakobi, M., and Maier, J. P. 1993. "Electronic and Infrared Spectra of C_{60}^{+} and C_{60}^{-} in Neon and Argon Matrices." *Chem. Phys. Lett.* **211**, 227–34.
Kondo, H., Momose, T., and Shida, T. 1995. "Reinvestigation of the Lowlying Electronic States of C_{60}^{-}." *Chem. Phys. Lett.* **237**, 111–14.
Wigner, E. P. 1959. *Group Theory and Its Application to the Quantum Mechanics of Atomic Spectra.* New York: Academic Press.

Index